Taxing Automobile Emissions for Pollution Control

NEW HORIZONS IN ENVIRONMENTAL ECONOMICS

General Editor: Wallace E. Oates, *Professor of Economics, University of Maryland*

This important series is designed to make a significant contribution to the development of the principles and practices of environmental economics. It includes both theoretical and empirical work. International in scope, it addresses issues of current and future concern in both East and West and in developed and developing countries.

The main purpose of the series is to create a forum for the publication of high quality work and to show how economic analysis can make a contribution to understanding and resolving the environmental problems confronting the world in the late twentieth century.

Recent titles in the series include:

Economics for Environmental Policy in Transition Economies
An Analysis of the Hungarian Experience
Edited by Péter Kaderják and John Powell

Controlling Pollution in Transition Economies
Theories and Methods
Edited by Randall Bluffstone and Bruce A. Larson

Environments and Technology in the Former USSR
Malcolm R. Hill

Pollution and the Firm
Robert E. Kohn

Climate Change, Transport and Environmental Policy
Empirical Applications in a Federal System
Edited by Stef Proost and John B. Braden

The Economics of Energy Policy in China
Implications for Global Climate Change
ZhongXiang Zhang

Advanced Principles in Environmental Policy
Anastasios Xepapadeas

Taxing Automobile Emissions for Pollution Control
Maureen Sevigny

Global Environmental Change and Agriculture
Assessing the Impacts
Edited by George Frisvold and Betsey Kuhn

Fiscal Policy and Environmental Welfare
Modelling Interjurisdictional Competition
Thorsten Bayindir-Upmann

The International Yearbook of Environmental and Resource Economics
1998/1999
A Survey of Current Issues
Edited by Tom Tietenberg and Henk Folmer

The Economic Approach to Environmental Policy
The Selected Essays of A. Myrick Freeman III
A. Myrick Freeman III

Taxing Automobile Emissions for Pollution Control

Maureen Sevigny
Oregon Institute of Technology
Oregon, United States

NEW HORIZONS IN ENVIRONMENTAL ECONOMICS

Edward Elgar
Cheltenham, UK • Northampton, MA, USA

Published by
Edward Elgar Publishing Limited
8 Lansdown Place
Cheltenham
Glos GL50 2HU
UK

Edward Elgar Publishing, Inc.
6 Market Street
Northampton
Massachusetts 01060
USA

A catalogue record for this book
is available from the British Library

Library of Congress Cataloguing in Publication Data

Sevigny, Maureen, 1955–
        Taxing automobile emissions for pollution control / Maureen
Sevigny.
                (New horizons in environmental economics series)
        Includes bibliographical references.
        1. Environmental impact charges—California, Southern. 2. Motor
vehicles—Motors—Exhaust gas—Environmental aspects—California,
Southern. 3. Pollution—Government policy—California, Southern.
I. Title. II. Series: New horizons in environmental economics.
HJ5327.S48 1998
336.2—dc21   97–44510
                                                                CIP

ISBN 1 85898 767 9

Printed and bound in Great Britain by
MPG Books Ltd, Bodmin, Cornwall

To dad

Contents

List of Figures

List of Tables

Acknowledgements

Many people have contributed generously to this study and it is one of life's simple pleasures to say 'thank you' for their help.

Arlee T. Reno of Cambridge Systematics, Inc. introduced me to the question of how to model motorists' responses to a tax on automobile emissions and encouraged me to pursue further research in this area. Arlee's keen insights and knowledge, tempered with his easy laughter, made it a genuine pleasure to work with him.

Wallace E. Oates has guided my efforts for several years, first as my dissertation advisor at the University of Maryland and, more recently, as the general editor of this series. Wally has given generously of his time and expertise throughout our association. His comments have always been thoughtful, yet gentle, as he has tried to guide me towards the highest standards of excellence. I am extremely grateful.

John Ritter, my colleague at Oregon Institute of Technology, has enhanced my understanding of atmospheric chemistry while helping me see environmental issues through a more integrated, interdisciplinary perspective. John's vision and counsel have helped far more than he realizes. I also appreciate John's assistance with proofreading and his suggestions regarding the specification of several equations in an earlier draft of this work.

Mark Carlock of the California Air Resources Board, Chris Kavalec of the California Energy Commission, Dave Brzezinski of the US Environmental Protection Agency, Gale Newton of the Illinois Environmental Protection Agency and Robert Anderson of the American Petroleum Institute deserve special thanks for providing much of the data used in this study.

Finally, my husband Art, who has graciously accepted the many hours this endeavour has required, has my heartfelt gratitude. Thank you, Art, for all that you add to my life. Without you, it would not be a balanced equation.

Preface

Controlling emissions from mobile sources has proven to be one of the most challenging dimensions of air quality management. Moreover, it has become clear that more effective control of mobile sources is essential if many urban areas are to achieve the mandated air quality standards. Motor vehicles are primary sources of carbon monoxide and the various pollutants that combine to produce tropospheric ozone; without reduced emissions of these pollutants, many urban areas have little hope of attaining the prescribed standards for air quality.

Environmental agencies have, in the main, resorted to various forms of direct control for the regulation of mobile sources. These measures include, first and foremost, the installation of emissions control devices on new motor vehicles. However, the effectiveness of such devices depends importantly on their use and on automobile maintenance and operation. To address these issues, authorities in many areas have introduced vehicle inspection systems that attempt to identify vehicles with excessive emissions and to ensure that the requisite measures are taken to correct the deficiencies. Such systems have, however, achieved only mixed success; in addition, they are costly (including time costs) and often the source of much annoyance to vehicle owners.

It remains the case that a relatively small fraction of vehicles on the road accounts for a disproportionately large share of pollutant emissions. An effective regulatory measure would identify these sources and introduce a system of incentives to clean them up. From this perspective, a system of emissions taxes is very appealing. If we could accurately monitor the emissions from each vehicle in an air quality-compromised area, we could introduce a tax per unit of emissions that would weigh heavily on those vehicles that are significant polluters. Such a penalty on emissions could establish a powerful incentive, either for the scrappage (or reduced use) of such vehicles or for repair efforts to reduce their emission levels.

In this volume, Maureen Sevigny provides an important and seminal study of such a system, using the California South Coast Air Basin as a case study. She first constructs basic conceptual models of automobile ownership and driving behaviour; the models are themselves valuable contributions. Then, having assembled a rich body of data from Southern California and a variety of other sources, Sevigny is able to make her models operational and to simulate the effects of a tax on the polluting emissions from cars in the South Coast region. The study is helpful in many different ways. First, the design of such a tax itself presents some difficult issues. What exactly is to be taxed and how? Sevigny carefully considers the range of issues in the design of such a system of emissions taxes and then proposes an operational form of a set of such taxes. Next, she turns to her model to simulate the effects of the taxes on vehicle ownership, usage and emissions rates.

The findings are impressive. The taxes induce large reductions (in some cases well over 50 per cent) in emissions of the various pollutants. And this results from a variety of responses on the part of owners and drivers. It would no longer pay to keep many older and heavily polluting cars; they would simply be scrapped. In addition, it would now pay to keep cars in good operating condition so as to reduce rates of pollutant emissions and to alter driving behaviour so as to reduce the number of miles driven. As Sevigny's work makes clear, emissions taxes on automobiles are potentially very powerful regulatory instruments. They effectively target emissions themselves, not a surrogate such as levels of gasoline consumption (as under a gas tax) and thereby provide a direct incentive for adjusting behaviour in many different ways so as to reduce polluting emissions.

Sevigny's study also provides some intriguing results concerning the fiscal aspects of emissions taxes on mobile sources. Such taxes are a large potential source of revenues; Sevigny estimates that her proposed tax could raise over $1 billion in the South Coast region. In exploring the range of properties of such a tax, she finds (not surprisingly) that these taxes are quite regressive in their pattern of incidence. This suggests that were a mobile source emissions tax to be introduced in a revenue-neutral fashion, it might well be desirable to reduce other taxes in such a way as to offset some of the regressivity of the emissions tax.

The actual monitoring of mobile source emissions on an individual and regular basis presents a formidable challenge to the proposed tax. However, as Sevigny suggests, this hardly seems beyond our capacities. With continuing progress in the technology of monitoring, it is far from unimaginable that remote-sensing devices could become available to measure reasonably accurately the levels of all pollutants coming from the tailpipes of individual vehicles, much as carbon monoxide can be measured today. Even in the absence of remote sensors, Sevigny argues convincingly that emissions could be estimated for each vehicle based on a periodic emissions test and the number of miles travelled.

The Sevigny study is important, for it makes clear the tremendous potential of emissions taxes for the regulation of mobile source emissions. Such taxes could induce major changes in ownership and driving behaviour that would result in large reductions in emissions. Moreover, these reductions would be achieved in a relatively cost-effective way. Over the longer haul, such taxes would also provide a powerful incentive for the development of cleaner motor vehicles. We need, I believe, to think again about the role of emissions taxes for mobile sources; Sevigny's work suggests that economic incentives have much to offer in our efforts to control pollutants from automobiles.

Wallace E. Oates
Professor of Economics, University of Maryland,
and University Fellow, Resources for the Future

1. Introduction

The goal is clean air, at the lowest possible cost.[1]

The federal Clean Air Act, as amended in 1970, charged the US Environmental Protection Agency (USEPA) with establishing the National Ambient Air Quality Standards (NAAQS) for six pollutants, including carbon monoxide (CO) and ozone.[2] The standards were to be based on health criteria (protecting the weakest in the population from harm) without regard for the cost of achieving those standards.

Regions which fail to meet the standards are designated as non-attainment areas and are required by the 1990 Clean Air Act Amendments (CAAA) to develop and implement plans to reduce emissions. California's South Coast Air Basin[3] is a serious non-attainment area for CO and the nation's only extreme ozone non-attainment area. According to Lents and Kelly (1993), the South Coast Air Basin exceeded one or more of the NAAQS on 184 days in 1991.

Transportation is a significant source of CO and ozone precursors[4] in the South Coast region. Jack Faucett Associates (1992) estimated that on-road mobile sources, including passenger cars, light and heavy trucks, vans, buses and motorcycles, produced 46 per cent of hydrocarbon (HC), 59 per cent of nitrogen oxides (NO_x) and 87 per cent of CO emissions in the South Coast region in 1985. The California Environmental Protection Agency (1993) estimated that over 90 per cent of CO and about 60 per cent of ozone precursors were produced by mobile sources. The South Coast Air Quality Management District (SCAQMD, 1991) estimated that mobile source HC emissions must be reduced by 84 per cent if the NAAQS are to be attained by 2010. The inescapable conclusion is that the South Coast region *must* reduce mobile source emissions if the region is to achieve the national standards. The question is, what is the most efficient way to accomplish this?

ATTAINMENT: NOT BY TECHNOLOGY ALONE

Mobile source abatement can result from improved vehicle maintenance, retrofitting older vehicles with newer emissions control technology (Ely, 1994), driving fewer miles, driving under more optimal conditions[5] and replacing existing vehicles with lower-emitting vehicles.   Current and foreseeable technologies will be unable to produce the required emissions reductions for the South Coast. As Zimmer (1994) explains:

> The Clean Air Act Amendments of 1990 (CAAA) requires areas not in attainment of air quality standards to adopt measures to reduce emissions, many of which relate to transportation sources. A majority of these measures are aimed at modifications to the technologies associated with automobile engines or fuels. The cumulative effect of the required measures will not be sufficient to enable many urban areas to attain National Ambient Air Quality standards by the CAAA deadlines. The CAAA requires those areas to implement other measures in order to sufficiently reduce emissions.

Zimmer's belief that technology-based abatement will be inadequate to achieve the air quality standards is consistent with the SCAQMD view that both regulatory solutions and behavioural adjustments will be needed to reach attainment (SCAQMD, 1991, p. 4–38).

Because the CAAA gives non-attainment areas some discretion as to which source(s) should abate in what way(s),[6] and because technological solutions are unlikely to achieve sufficient emissions reduction, emissions pricing is viewed as a potentially attractive component of SCAQMD's plan. While the list of 'command and control' mobile and stationary source regulations is long and detailed (SCAQMD, 1991, Chapter 7), the 1991 *Air Quality Management Plan* also includes provisions to consider emissions fees, congestion fees, toll roads, tax credits and marketable permits to 'complement the existing regulatory system' (SCAQMD, 1991, p. 4–38). The goal, after all, is 'clean air, at the lowest possible cost' (SCAQMD, 1991, p. ES–18).

LIMITATIONS OF REGULATIONS AND TRADITIONAL TRANSPORTATION CONTROL MEASURES

Historically, most pollution control has taken the form of regulation, a 'command and control' approach to environmental policy. Regulations range from requiring the use of certain technologies, prohibiting or restricting the use of certain equipment or processes, and requiring particular forms of equipment inspection and monitoring, to restricting or prohibiting the output of specific pollutants. Most regulations were designed with little regard for the cost of abatement.[7]

In their joint report, *Clean Air Through Transportation* (1993), the US Department of Transportation (USDOT) and the USEPA claimed that traditional transportation control measures (TCMs) aimed at decreasing motor vehicle trips, vehicle miles of travel (VMT) and congestion by encouraging off-peak travel or use of transportation modes other than the single-occupancy motor vehicle (SOV), have 'not generated significant air quality benefits' (USDOT/USEPA, 1993, p. 2).

Among the 16 TCMs listed in Section 108(f)(1) of the CAAA,[8] only the employer-based trip reduction/employee commute option is mandatory, and only for large employers in serious CO and severe and extreme ozone non-attainment areas, such as the South Coast region. Many of the Section 108(f)(1) TCMs focus on the supply of alternatives to the SOV, such as public transit, traffic flow improvements, high-occupancy vehicle travel lanes and non-motorized vehicle paths and accommodations. USDOT/USEPA (1993, p. 9) claimed the 'regulatory TCMs included in Section 108(f)(1) and others which place restrictions on automobile travel show greater emissions reduction potential than TCMs that increase the supply of transportation alternatives' but, 'preliminary indications from around the country indicate that traditional TCMs will yield only a 1 to 2 per cent reduction in mobile source emissions, far short of what some areas need'.

The USDOT/USEPA report suggests that 'economic/market-based' TCMs which affect the demand for SOV travel may be more promising for emissions reduction than the regulatory or traditional transportation supply-oriented TCMs (USDOT/USEPA, 1993, pp. 3, 9–10). The incentives, including congestion pricing, higher parking prices and emissions charges, would increase the cost of SOV travel and, in the process, enhance the relative attractiveness of alternatives. Many individuals would choose to continue their current travel patterns, but others would be induced to switch to alternatives that would result in emissions reduction.

EMISSIONS TAXATION: A POSSIBLE, COST-EFFECTIVE APPROACH

Market-based incentives have a strong theoretical foundation but only a short history of actual use in pollution control. Incentives range from taxes per unit emitted, to subsidies for each unit abated, to a tradable permit system in which the polluter must acquire a permit for each unit emitted.

Under a market-based approach, each potential emitter must decide if it is worth paying the price to pollute or if abatement is more cost-effective. The individual decides how much to abate and by what means. Abatement may result from adopting new technologies or processes, overhauling existing facilities or operating procedures, or shutting down the polluting activity altogether.

Baumol and Oates (1988) have shown that a tax per unit of emissions is a cost-effective tool for emissions reduction because the tax ensures that each source abates up to the point where the unit tax equals its marginal abatement cost (MAC). As Figure 1.1 shows, a polluter would choose to reduce emissions by Q^* units when faced with the unit tax, given the polluter's MAC curve.

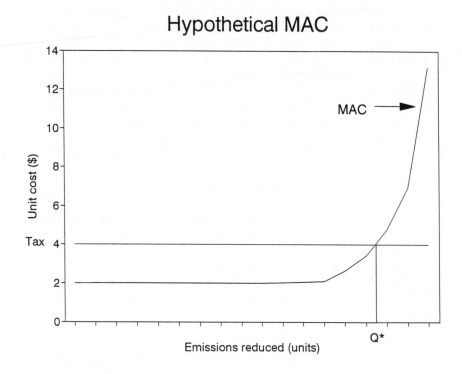

Figure 1.1 MACs and emissions abatement

Polluters with different MAC curves would choose different Q^*s. Polluters with low MACs would abate higher quantities than would polluters with high MACs. The net result is an efficient outcome in which area-wide abatement, the sum of the individual Q^*s, is achieved at minimum cost.

If the total quantity abated equates the marginal social benefit of abatement with the marginal social cost, the result would also be optimal. Optimality would be much harder to achieve than efficiency because of the difficulties involved in measuring the marginal social benefit and marginal social cost of abatement.[9]

SCOPE OF THIS STUDY

This study examines the role that a mobile source emissions tax could play in reducing emissions in the South Coast Air Basin. Beginning with a theoretical discussion of a first best tax, a second best tax on passenger vehicles is developed and used to simulate the effects of the tax in the South Coast Air Basin for 1990–91. The study focuses on vehicle owners' responses to the tax which would lead to emissions reduction. This study contains detailed analyses of:

- the design of a mobile source emissions tax;
- behavioural responses that lead to emissions reductions, including reductions in the household's VMT and the scrappage of low-value, high-emitting vehicles;
- emissions reduction that would result from the tax;
- the effects of the tax on households in different income quintiles.

The study uses a simulation model to analyse the sensitivity of travel demand (and the resulting emissions) to different tax rates and demand elasticities. The TIERS model (Tax-Induced Emissions Reduction Simulation) shows that, in the absence of vehicle tampering or tax evasion, the emissions tax has the potential to induce significant reductions in emissions from household vehicles, even when travel demand is relatively inelastic with respect to price. The mobile source emissions tax is shown to induce reductions in HCs, CO, and NO_x that represent as much as 47, 72 and 62 per cent respectively of the 1994 Basin-wide reduction targets from all sources. The TIERS model also shows that the emissions-reducing potential of a change in gasoline taxes is considerably less than that of an emissions tax.

One of the difficulties encountered in this study was the lack of a single household-level dataset containing information on vehicle ownership, vehicle usage, vehicle emissions rates, other vehicle characteristics (including operating costs) and household demographic variables. In response to this data shortcoming, the study has integrated several diverse datasets. The data compilation and analysis may well be one of the major contributions of this work. Appendix C contains a thorough discussion of the data sources and their applicability to the South Coast Air Basin.

NOTES

1. South Coast Air Quality Management District (1991) p. ES–18.
2. The NAAQS established an eight-hour concentration of 9 parts per million (ppm) for CO and a one-hour concentration of 0.12 ppm for ozone.

3. The South Coast Air Basin includes Orange County and the non-desert portions of Los Angeles, Riverside and San Bernardino counties.
4. Ozone is produced by a photochemical reaction of NO_x and reactive HCs. The technical literature distinguishes among various classes of hydrocarbons, including volatile organic compounds, reactive organic gases (ROG), non-methane hydrocarbons and non-methane organic compounds. While these are non-trivial differences, I am referring to these compounds by the broader term of 'hydrocarbons', consistent with Federal Highway Administration (1992a) usage.
5. See Chapter 2 for a discussion of the effects of vehicle speed, acceleration and ambient temperature on vehicle emissions.
6. See Appendix A for a list of specific CAAA requirements for CO and ozone non-attainment areas.
7. See Portney (1991) for a good discussion of the cost-ineffectiveness of most regulations.
8. Appendix B lists and describes Section 108(f)(1) TCMs.
9. Damage caused by a unit of HC, CO or NO_x emissions varies both temporally and spatially and is also affected by wind direction and speed, ambient temperature and topography. The relative ambient concentrations of HC and NO_x determine whether additional NO_x emissions act as a sink or source of ozone. See Milford et al. (1989) and Findlayson-Pitts and Pitts (1993) for a discussion of ozone photochemistry.

2. Designing a Tax on Mobile Source Emissions

The use of a tax per unit of emissions is well developed in the externalities literature,[1] although the focus traditionally has been on stationary sources. However, once a unit of emissions reaches a receptor site, it is completely fungible with a unit emitted from any other source. All pollution sources that have similar effects on a receptor in a given airshed should be treated similarly. There is no theoretical reason to refrain from extending the unit tax theory to mobile sources.

There are several questions to address when designing an emissions tax:

- What is the taxable unit?
- How is the taxable emission monitored?
- What is the appropriate unit tax rate?
- What are the institutional arrangements needed for assessment, collection and monitoring compliance?

THE RELATIONSHIP BETWEEN EMISSIONS AND AIR QUALITY

One of the most difficult issues to resolve is the relationship between an emitted pollutant and its resulting effect on ambient air quality. Emissions in a given location may or may not degrade ambient air quality. Temperature, wind, sunshine and numerous other topographic and meteorological variables affect the process:

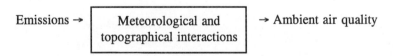

This relationship is especially complex for ozone production, which can be enhanced or inhibited by the relative ambient concentrations of HC and NO_x (Schere, 1988; National Research Council, 1991).[2]

A theoretical first best mobile source emissions tax could be designed as a cost-effective means of achieving either the NAAQS *or* a targeted reduction in emissions. If the target is the attainment of the NAAQS, the analysis must address the process by which emissions lead to ambient air quality degradation. If the target is a specific quantity of emissions reduction, the analysis is far simpler because the 'interactions' box depicted above is not relevant.

Analysis of CO is straightforward because the ambient concentration is proportional to emissions. An emissions reduction policy will simultaneously target attainment of the NAAQS.

This simple relationship does not hold for HC and NO_x. A tax designed to reduce ozone would have to account for the interactions which convert HC and NO_x to ozone at one or more receptors in the airshed. Because there is a delay between the time of emissions and the formation of ozone, it would be necessary to know whether ozone-conducive conditions would be present when the HC or NO_x reached its destination.

DESIGNING A FIRST BEST TAX ON MOBILE SOURCE EMISSIONS

Assuming it were possible to monitor emissions from each vehicle, the first best tax would be based on the effect each unit of HC, CO and NO_x emissions had on the ambient air quality in the Air Basin.

Consider M vehicles, each emitting three pollutants (HC, CO, NO_x), and K receptors in the airshed. The effect on individual receptor k from vehicle m can be described by D_{mik}, the degradation in air quality caused by the ith pollutant.

The damage caused by CO emissions, D_{mik}, i=grams of CO, would be directly related to the concentration of i at receptor k. The damage caused by HC emissions, D_{mik}, i=grams of HC, would be a function of the transformation of HC into ozone. This in turn would depend on the HC/NO_x ratio at receptor k as well as atmospheric conditions. The damage caused by NO_x, D_{mik}, i=grams of NO_x, would depend on the same factors as HC. If ozone did not result from the HC or NO_x emission, the damage would be zero.[3]

Since emissions of HC or NO_x can take several hours to reach a given receptor and produce ozone, emissions must be measured as:

E_{mijt},	$m = 1,..,M$	individual vehicle
	$i = $ HC, CO, NO_x	pollutant emitted (grams)
	$j = 1,...,J$	emission location
	$t = 1,...,T$	time of emission

Assuming E_{mijt} and D_{mik} could be measured, it would be possible to set a tax rate for each gram of pollutant emitted, reflecting the degradation it caused at any receptor it affected. Rates at one receptor need not equal those at another, since the level of degradation need not be the same across receptors.[4]

In a first best world, an on-board sensor would record the emissions of CO, HC and NO_x. Global positioning system technology would pinpoint vehicle location and an on-board computer would calculate D_{mik}, accounting for atmospheric conditions and topography. A meter would record the ongoing D_{mik}, much as a taxi meter records fares. The emissions meter would be cleared periodically by payment of the accrued fee.

If the first best tax were designed to target emissions reduction instead of ambient air quality it would be necessary to measure emissions as well as establish an appropriate tax rate for each pollutant. The appropriate tax rate is the one which would achieve the desired emissions reduction. This would likely be determined through trial and error, assuming the political process permitted the rate to be altered as needed, because the individual MAC curves would not be known with certainty. If the MACs were known, the first best tax rate would be set such that the sum of each emitter's optimal abatement equalled the emissions reduction target, as shown in Chapter 1.

DESIGNING A SECOND BEST TAX ON MOBILE SOURCE EMISSIONS

A first best tax requires the ability to measure each unit of emissions and assess the effect of each pollutant on ambient air quality. However, the air quality effect of a given unit of emissions is uncertain and, with over eight million passenger cars in the South Coast region, it is impractical to retrofit each with a continuous emissions monitor.[5] It is possible, instead, to estimate a vehicle's annual emissions of each pollutant, i, as the product of its measured emissions per mile (as determined by an I/M 240 test at a vehicle inspection station) and its annual VMT:

$$E_i = \text{emissions}_i/\text{mile} * VMT \qquad i=\text{HC, CO, NO}_x$$

E_i is only an approximation of the annual emissions of the ith pollutant because emissions vary with vehicle usage and several exogenous factors:

$$E_{mijt} = f(\text{vehicle speed, acceleration/deceleration, maintenance,}$$
$$\text{engine starts, air temperature, gasoline volatility and}$$
$$\text{additives, evaporation, refuelling})$$

Travel velocity has a strong non-linear inverse relationship with emissions of

HC and CO, especially at speeds below 20 miles per hour.[6] Acceleration and deceleration produce higher emissions per mile than driving at a steady speed. HC emissions are especially high when an engine is started with a cold catalytic converter (cold start emissions). Some HC evaporates during engine cool down (hot soak emissions) after a running engine is turned off. Some HC also escapes during refuelling. The number of engine starts and the frequency and duration of acceleration and deceleration cannot be measured by the I/M 240 vehicle inspection's dynamometer, although it is designed to simulate driving under different speeds and load conditions. CO emissions are inversely related to ambient temperature, with most violations of the NAAQS occurring below 50°F.[7]

The variability of on-road emissions suggests that the emissions rate calculated by the I/M 240 inspection risks serious error in measuring a vehicle's true emissions. This is not a problem for the second best tax, however, because it relies on *average* emissions per mile to estimate total emissions. The I/M 240 test does a reasonable job of simulating average driving conditions. It will also distinguish high emitters from low ones, allowing the tax to be greater for the former.

Annual mileage (VMT) can be measured as the change in the vehicle's odometer over a given time interval. There is serious concern about potential odometer tampering, but that is an enforcement problem, not a design issue. From the design standpoint, it is sufficient to know that each vehicle's VMT can be measured.

The second best tax is designed to target emissions reduction, not the direct attainment of the NAAQS. This is consistent with the CAAA, which requires the South Coast Air Basin to reduce HC emissions from all sources by 15 per cent from the 1990 base level by 1996 and to reduce emissions an additional 3 per cent per year until the NAAQS are attained. The second best tax is designed as a cost-effective approach to meeting this mandate.

The second best tax, to be assessed annually on each household vehicle in the South Coast Air Basin, is measured as:

$$Annual\ tax = (a*EPM_1 + b*EPM_2 + c*EPM_3)*VMT$$

where:

a, b and c are the tax rates per gram of HC, CO and NO_x emissions, respectively;
EPM_1 is the measured HC emissions rate (g/mile);
EPM_2 is the measured CO emissions rate (g/mile);
EPM_3 is the measured NO_x emissions rate (g/mile);
VMT is annual mileage, measured as the change in odometer readings.

Basing the tax on each vehicle's total emissions gives drivers the flexibility to choose their preferred method of abatement: reduce mileage, reduce emissions per mile, or both. Drivers will face the same tax rate per gram of pollutant i, but different tax rates per mile, as emissions per mile vary from car to car.

The mobile source emissions tax does not address the congestion externality that driving often generates, nor is it designed as a pure revenue measure. It is designed to reduce emissions. In the process, it will raise revenues and, possibly, reduce congestion. Revenue estimates are discussed in Chapter 4 and the congestion issue is discussed briefly in Chapter 6.

One of the more interesting questions is how to determine the tax rates a, b and c. Tax rates could be set so as to induce the desired level of abatement for each pollutant. If this approach is used, c should be set to zero because there is no mandatory NO_x reduction target for the South Coast. HC and CO tax rates would depend on the elasticity of emissions with respect to the tax. Key elements of this would include the elasticity of VMT with respect to the increased cost per mile of travel and the extent of emissions reduction from increased maintenance and scrappage of high emitters. This process would be iterative because the elasticities would not be known *a priori*. It would also be necessary to establish emissions reduction targets for mobile sources; current mandated targets are for area-wide reductions regardless of the source.

A second approach would be to set the tax rates equal to the lowest alternative control cost in the South Coast region for each pollutant. In the absence of transactions costs, the mobile source emissions tax would then approximate a local least cost solution. The marginal abatement cost for mobile source emissions would be equal to the lowest marginal abatement cost for stationary or area sources, providing all emitters in the South Coast with the same incentive to abate.

Estimates of marginal emissions control costs have been obtained from several sources. The US Congress Office of Technology Assessment (OTA) (1992) estimated the value of emissions reduction (cost-effectiveness based on alternative control costs) on a nation-wide basis. The California EPA (1993) estimated the cost per ton of HC and NO_x abatement from early retirement of old cars in California. Anderson (1990) estimated ranges of stationary source control costs in the South Coast region. Wang and Santini (1994) estimated the alternative control costs in the South Coast region based on a study from the California Energy Commission. Table 2.1 summarizes the findings.

Wang and Santini's figures appear to represent the control costs of electric utilities in the South Coast and are not necessarily the minimum control costs for the region. Both Anderson and OTA represent minimum control costs.

The mobile source emissions tax rates must be set equal to the minimum alternative control costs in the region for the tax to be efficient. It is not surprising that the marginal abatement costs in the South Coast exceed the national figures because emissions controls have been in place for some time in the region. Many sources have already undertaken the low-cost abatement projects and are now facing a steeper portion of their MAC curves. There is no expectation that the tax rates would be optimal, although it is possible that they could be.

Table 2.1 Four estimates of pollution control costs ($/ton)

	OTA	California EPA	Anderson	Wang and Santini
HC	3050	4 700	15 000	18 900
CO	300	N/A	200	9 300
NO_x	2750	18 000	8 000	26 400

For this study, the tax rates chosen are based on the control costs derived by Anderson (1990) for stationary sources in the South Coast: $15000 per ton of HC, $200 per ton of CO and $8000 per ton of NO_x. The rates are somewhat arbitrary because true control costs of sources are generally not known. Anderson's estimates have been chosen because they are based exclusively on the South Coast region.

The control costs need not be adjusted to reflect the additional cost imposed on vehicle owners when they take their vehicles to an emissions inspection station because vehicle inspection is already required under current law.[8]

SUPERIORITY OF THE SECOND BEST TAX TO A GASOLINE TAX

The second best tax is superior to a gasoline tax for reducing emissions. While a gasoline tax encourages the purchase of fuel-efficient cars, the correlation between emissions per mile and miles per gallon (MPG) is weak. As Figure 2.1 shows, in a pilot study by the Illinois EPA which targeted and scrapped high-emitting vehicles, there was a wide range of HC emissions for any given MPG among the vehicles for which both readings were available.

This result should not be surprising, given the physical properties that determine fuel use and emissions. Fuel efficiency depends in part on vehicle weight, engine size and performance, vehicle aerodynamics and gearing. These features generally do not affect emissions, which depend on the completeness of the fuel burn and the presence or absence of emissions control systems such as a catalytic converter.

HC emissions vs MPG
181 vehicles: ILEPA

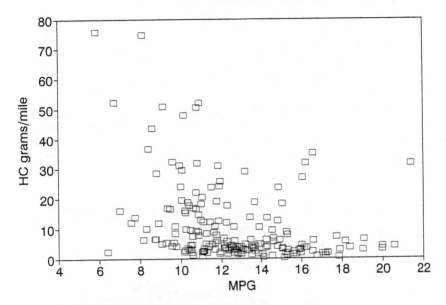

Figure 2.1 Weak correlation of HC emissions and MPG

The weak correlation of emissions and MPG also arises, in part, from the way the control standards were designed. Emissions standards are established for each model year. All cars built in a given year are designed to emit at or below the same rate. Fuel efficiency is allowed to vary across the makes and models in the model year; the requirements are only that the fleet average MPG does not fall below a certain level. Thus a 1984 car designed to emit 0.41 grams of HC/mile and achieving 20 MPG will emit 8.2 grams of HC/gallon of gasoline, while another 1984 car emitting 0.41 grams of HC/mile and achieving 50 MPG will emit 20.5 grams of HC/gallon. Increasing the price per gallon does not provide a proportional incentive to reduce emissions.

A gasoline tax increase may also *increase* VMT in the long run. If a less fuel-efficient vehicle is replaced by a more fuel-efficient one, the reduced operating cost of the new vehicle will encourage increased VMT. Depending upon the emissions rates of the old and new vehicles, the increased VMT could lead to an increase in total emissions. This is clearly the wrong incentive for long-run emissions reduction.

SUPERIORITY OF THE SECOND BEST TAX TO A MODEL YEAR TAX

A model year indexed registration surcharge is often suggested as an incentive to remove older vehicles from the fleet and remove a large percentage of fleet emissions. *On average*, older vehicles do emit at a higher rate than newer ones. Older vehicles were designed to meet less stringent emissions standards, and all emissions control systems deteriorate with age. However, a model year surcharge is inefficient because there is a wide distribution of emissions among vehicles of the same age.

As Figure 2.2 shows, the cleanest 20 per cent of vehicles in a model year produce about 5 per cent of that year's HC grams/mile, while the dirtiest 20 per cent produce over half. A model year indexed registration fee ignores the variation of vehicle emissions rates within that model year.

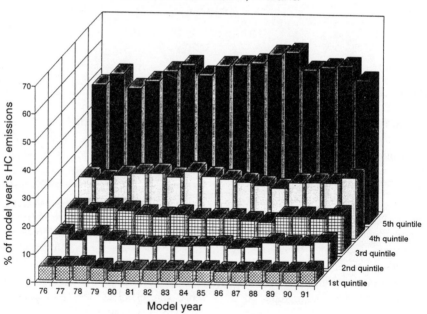

Figure 2.2 HC distribution within model years

VMT also varies widely within model years, as Figure 2.3 shows. A high-emitting vehicle that travels very few miles can produce fewer grams of emissions than a low-emitting vehicle that is driven many miles. Since it is the product of emissions per mile and miles travelled that determines total emissions, it is important to target both the emissions rate and VMT.

A model year surcharge ignores variation in emissions per mile and VMT, and treats all vehicles of a given age alike. This provides no incentive for owners of dirtier vehicles to reduce emissions, either by reducing their emissions per mile or driving fewer miles, and penalizes owners of low-emitting or lightly travelled vehicles. This is not efficient.

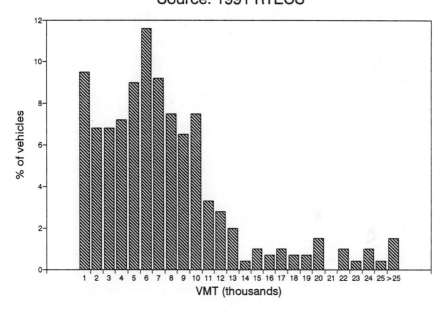

Figure 2.3 VMT distribution for older vehicles

NOTES

1. See, for example, Atkinson and Lewis (1974) and Baumol and Oates (1988).
2. The National Research Council (1991) study shows some regions in the South Coast Air Basin will suffer *increased* ozone if NO_x is reduced and HC left unchanged. Given the joint product nature of mobile source emissions, NO_x reductions are typically accompanied by HC reductions, although the proportions are highly variable. HC reduction is generally a necessary, but not sufficient, condition for ozone reduction. CO reduction is both necessary and sufficient for reducing the ambient level of CO. CO often concentrates near its source, creating a 'hotspot' problem.
3. This analysis focuses on NO_x solely as an ozone precursor. This is shortsighted because a federal ambient air standard exists for one component of NO_x, nitrogen dioxide (NO_2).
4. The National Research Council (1991) shows great variability in the HC/NO_x ratio across the South Coast Air Basin. A gram of NO_x emitted in central Los Angeles has a different ozone-forming potential than a gram emitted further east (for example, in Chino or San Bernardino).
5. The technology for continuous monitoring exists and has been used experimentally to study the effects of vehicle speed, acceleration and engine temperature on the rate of exhaust emissions. See, for example, Cadle et al. (1993).
6. Cambridge Systematics, Inc. (1991) estimated that a ten-mile trip taken at 20 mph produces 350 per cent more running exhaust HC than the same trip taken at 55 mph. Sigsby et al. (1987) showed that as speed declined from the Federal Test Procedure's 19.6 mph to 7.1 mph (not atypical for congested urban roads), HC emissions rose by 83 per cent.
7. Cambridge Systematics, Inc. (1991), p. 22.
8. California currently inspects passenger vehicles biannually. The CAAA require an annual inspection. California and the USEPA reached an agreement in 1994 permitting biannual inspections to continue.

3. Effects on Travel Demand and Maintenance

An emissions tax would increase the cost of driving in proportion to the vehicle's emissions of HCs, CO and NO_x. Although the taxable unit is the gram of emissions, the tax is a *de facto* mileage tax, with the rate per mile a direct function of the vehicle's emissions:

$$tax = \sum_i (tax\ rate/gram_i * grams/mile_i) * VMT \quad i=\text{HC, CO, NO}_x$$

When the vehicle emissions are measured and the tax rates for each pollutant are known, the tax rate per mile is readily calculated. For example, with the proposed tax rates of 1.65 cents/gram of HC, 0.022 cents/gram of CO and 0.88 cents/gram of NO_x, a vehicle emitting 1 gram of HC, 10 grams of CO and 2 grams of NO_x per mile would be assessed a tax of 1.65*1+0.022*10+0.88*2 =3.63 cents per mile.

By increasing the cost per mile in proportion to the emissions per mile, a vehicle owner would have an incentive to reduce the miles travelled and/or reduce the emissions per mile. Emissions per mile can be reduced through vehicle maintenance or, in the long run, by replacing the vehicle with a lower emitter. VMT can be reduced by forgoing trips, changing the route to a particular destination, or choosing an alternative mode of transportation such as public transit or walking. Each of the short-run responses will be discussed in this chapter. Vehicle turnover (scrappage) will be discussed in Chapter 5.

TRADITIONAL TRAVEL DEMAND MODELLING: THE FOUR-STEP PROCESS

Traditional transportation demand modelling consists of a four-step process: trip generation, trip distribution, mode split and trip assignment. This four-step process considers the factors which give rise to the decision to travel (trip generation), the decision of where to travel (trip distribution), the decision of how to travel (mode split) and the decision of which route to travel (trip

17

assignment).   Traditional travel demand models evaluate these four choices sequentially.

Trip generation models estimate the total number of trips entering or leaving a given land area (zone) over a specific period of time.   Trips are generally characterized in terms of their place of origin (home-based or non-home-based), and each type of trip is modelled separately.   Each model uses socioeconomic, land-use and location measures as exogenous variables, reflecting the idea that travel is a derived demand based on activities in given locations.   Travel demand is implicitly assumed to be independent of the transportation services or systems. Time and pecuniary costs of travel are not considered to be factors in trip generation.   A typical specification for home-based trips to work would be:[1]

$$T_i = \beta_0 + \beta_1 H_i + \beta_2 C_i + \beta_3 W_i + \epsilon_i$$

where:

T_i = the number of work trips originating in zone i;
H_i = the number of households living in zone i;
C_i = the number of cars owned by households in zone i;
W_i = the number of workers living in zone i.

The second step in the four-step process is to estimate the distribution of trip destinations.   The classic trip distribution model is the gravity model, loosely adapted from Isaac Newton's Law of Gravitation (1686).   This law, which states that the force of attraction between two bodies is proportional to the mass of the two bodies and inversely related to the square of the distance between them, has been used to estimate trip movements (attractions) from one zone to another.[2] Mathematically, the gravity model takes the form:

$$T_{ij} = K P_i P_j / (R_{ij})^n$$

where:

T_{ij} = one-way trip from origin i to destination j;
P_i = population in zone i;
P_j = population in zone j;
R_{ij} = the distance between origin i and destination j;
K, n = constants; $1 < n < 2$ usually.

The distance variable can be a proxy for travel time, although travel time can vary with the mode of transportation.   The constant, K, a direct analogue to

Newton's gravitational constant, has been the subject of numerous studies. While gravity is a physical constant, the behavioural nature of travel suggests that K should vary with the household characteristics in the two zones. The dynamic nature of these household characteristics further suggests that the usefulness of gravity models for prediction is likely to diminish over time unless they are recalibrated. Models developed by Harvey and presented by Cameron (1991, pp. 81–98) show some promise in incorporating economic determinants into the trip generation and distribution decisions.

The third step in the four-step process is to model the choice of travel mode. The simplest mode choice models limit the choice set to private vehicle or public transit. Complex models may also include non-motorized modes (walking, bicycling), single-occupancy and multi-passenger private vehicles, bus, train and multimodal choices (such as walk to bus, drive to bus), often in a nested choice framework.

Mode choice models assume a traveller chooses the mode which maximizes his or her utility, given modal and traveller characteristics. Travel time and pecuniary cost are important determinants of the utility derived from a particular mode. Mode choice models generally use a multinomial logit framework to estimate the probability that a trip from a given origin to a given destination will use a particular mode:

$$P_m = \frac{\exp(U_m)}{\sum_{i=1}^{j} \exp(U_i)}$$

where:

P_m = the probability of choosing mode m from the j available modes;
U_m = the utility derived from choosing mode m;
U_i = the utility derived from choosing the ith mode, $i=1,...,j$.

In this specification, the unit of travel is trips, not miles of travel. Trip length affects the utility of each mode through the expected travel time and pecuniary costs.

The fourth of the four-step models, route choice, assumes the traveller chooses the route which minimizes travel time. Travel time is a function of travel distance and speed; speed is a function of route characteristics (such as posted speed limit, presence or absence of traffic lights and passing lanes) and the volume of traffic. Route choice is modelled as a minimum-time-path network assignment problem and is solved in an iterative manner to account for the effect

of increased traffic assigned to a given route.   Iteration continues until all
travellers are assigned to routes that minimize travel time across the network,
given the assignments of all other travellers.

Route choice models do not maximize the utility of travel along a given route.
Route-specific factors, such as pavement condition, scenery and perceived
safety, are included only indirectly, to the extent that they affect traffic speed.

The four-step model was designed primarily for engineering studies of
roadway capacity and future requirements.   The sequential nature of the
solution, without feedback loops, and the lack of behavioural modelling in all
but mode choice make the four-step framework inappropriate for assessing the
effect of an emissions tax on household travel behaviour.

Since the tax would only affect the cost of personal vehicle usage, it is more
appropriate to analyse the tax in the context of the household's demand for miles
of travel in its own vehicles. Although this ignores the possibility of substitution
among modes, as well as the trip generation process and decision, there are
compelling reasons to use this approach.

HOUSEHOLD DEMAND FOR TRAVEL BY PERSONAL VEHICLE

Private vehicles are the overwhelming mode of choice for personal travel. As
Table 3.1 indicates, private automobiles and trucks accounted for more than 90
per cent of person-trips in each of the four years of the *Nationwide Personal
Transportation Survey (NPTS)*.[3] Since Table 3.1 includes households owning
no vehicles, the preference for the personal vehicle mode is even greater among
vehicle-owning households.

Table 3.1 Percentage of all person-trips by mode

	1969	1977	1983	1990
Automobile	85.1%	82.5%	81.5%	82.6%
Truck	5.6%	9.7%	11.6%	11.8%
Total Private Vehicle	90.7%	92.2%	93.1%	94.4%

Source: NPTS, Figure 14, p. 35.

In the 1990 *NPTS*, public transit use accounted for only 2.0 per cent of all
person-trips, down from 2.2 per cent in 1983 and 2.4 per cent in 1977 (*NPTS*,
p. 17). Work trips accounted for 40.6 per cent of all public transit trips in the
1990 survey, followed by school/church (21.8 per cent) and family/personal

(20.4 per cent) (*NPTS*, Figure 7, p. 18).

In central cities, public transit was chosen for fewer than 4 per cent of all person-trips in 1990 and about 10.5 per cent of all work travel, down from 4.8 and 11 per cent, respectively in 1985. Suburban public transit's share of all trips fell from just under 2 per cent to 1.2 per cent, and work trips fell from 3 per cent to just over 2 per cent (*NPTS*, Figures 9 and 10, pp. 20–21).

Even among low-income households, public transit is generally not the preferred mode. Among working households classified by the US Bureau of the Census as being in poverty, public transit was used for fewer than 10 per cent of work trips in 1989; driving alone accounted for nearly 60 per cent and carpooling approximately 18 per cent of all work trips (*NPTS*, Figure 18, p. 28).

The data strongly suggest that the decision to travel simultaneously implies the mode choice of private vehicle for most households. Instead of using the four-step model, it is more appropriate in the context of an emissions tax to consider the two choices facing the vehicle-owning household which has decided to travel: route choice and, possibly, the choice of which vehicle to use.[4]

Route choice can affect VMT, especially in urban areas where alternative routes between origin and destination frequently exist. Route choice may also affect cruising speed and the frequency and duration of acceleration and deceleration which, in turn, increase the variability of the vehicle's emissions per mile. However, since the emissions tax under study is determined by the measured (I/M 240) emissions rate, speed and acceleration will not be influenced by the tax. (If remote sensing were used to record emissions rates and assess the tax, speed and acceleration would become choice variables for the vehicle operator. Remote sensing is discussed briefly in Chapter 6.)

Households owning two or more vehicles also choose the particular vehicle to use for a given trip. To the extent that emissions rates vary across vehicles, the choice of which vehicle to operate affects trip emissions and the tax incurred. Since the emissions tax varies directly with the emissions rates of each vehicle, the multi-vehicle household would have an incentive to choose the lower-emitting vehicle for any given trip. This substitution effect may allow multi-vehicle households to reduce emissions without reducing travel.

EFFECTS OF THE EMISSIONS TAX ON VEHICLE MAINTENANCE

An emissions tax will encourage increased maintenance. Some maintenance will be a temporary fix – adjusted prior to the scheduled emissions inspection and unadjusted immediately afterwards – but some will be a permanent adjustment resulting in genuine emissions reduction.

Studies of emissions of on-road vehicles suggest that remote sensing, which automatically measures emissions as vehicles pass a roadside site, is needed to supplement or replace scheduled vehicle emissions inspections.[5] These studies argue that vehicle tampering, evasion of scheduled inspections and improper inspections allow many vehicles to emit at unacceptable rates. California expects to supplement its scheduled inspections with remote sensing, which should encourage motorists to perform proper maintenance and leave it in place.

There is a lack of comprehensive data on vehicle maintenance and few studies in this area. The dearth of information arises, in part, from the fact that maintenance is often a labour-intensive, non-market activity. Maintenance will become more important in the face of an emissions tax since any reduction in the emissions rate per mile will reduce the tax per mile. It is reasonable to expect that vehicle owners will perform all cost-effective maintenance, that is, all maintenance up to the point that the marginal cost of maintenance equals the marginal emissions tax reduction.

Simple, low-cost maintenance can often affect vehicle emissions rates significantly, particularly among older vehicles. Sommerville et al. (1987) reported that low-cost maintenance (generally below $50) could produce significant emissions reductions for many vehicles tested in California. Repairs ranged from replacing defective spark plugs and wires (a permanent fix) to adjusting the air/fuel ratio (potentially a temporary measure).

Table 3.2 shows a 33 per cent average reduction in HC emission rates across the vehicles in Sommerville et al.'s sample.

Table 3.2   HC emissions reduction from low-cost repairs

Vehicle vintage	HC g/mile *ex ante*	HC g/mile *ex post*	% HC reduction	Tax savings (cents/mile)	Break-even VMT at $50 repair
Pre-1975	10.49	6.59	37.2	6.44	776
1975–79	4.25	3.16	25.5	1.09	4587
Post-1979	1.89	1.29	31.4	0.60	8333
All	5.24	3.52	32.9	1.72	2907

Table 3.3 shows the magnitude of HC reduction in vehicles that attained at least a 50 per cent decline.

Tables 3.2 and 3.3 also show the break-even VMT where the tax savings equal a $50 maintenance cost. For example, a reduction of one gram per mile would save 1.65 cents per mile. The savings for 3030.3 miles would equal $50. Note that these tax savings and break-even calculations ignore reductions in CO and NO_x that would also result from the maintenance procedure. Tables 3.2 and 3.3 strongly suggest that vehicle owners would perform low-cost, emissions-reducing maintenance in response to the emissions tax, especially for high VMT vehicles.

Table 3.3 Repaired vehicles reducing HC emissions by 50 per cent or more

Vehicle vintage	HC g/mile *ex ante*	HC g/mile *ex post*	% HC reduction	Tax savings (cents/mile)	Break-even VMT at $50 repair
Pre-1975	21.99	8.00	63.6	13.99	357
1975–79	9.71	2.26	76.7	7.45	671
Post-1979	3.56	1.17	67.1	2.39	2092

EFFECTS OF THE EMISSIONS TAX ON VEHICLE USAGE

A household chooses the number and type(s) of vehicle(s) to own and the number of miles to drive each vehicle to maximize its utility. An emissions tax would lead to adjustments in vehicle holdings in the long run but, in the short run, the household's vehicle stock is fixed.[6]

Household vehicle usage (annual VMT of each vehicle) can be studied in the context of utility-maximizing behaviour, following the approaches of Train (1986) and Walls et al. (1993). The household's indirect utility function can be described as:

$$V_{nc_n} = f(Y, P_{nc_n}, X_{nc_n})$$

where:

n is the number of vehicles of type c_n owned by the household;
c_n is the class and vintage[7] for each of the n vehicles;
Y is the household's annual income;
P is a vector of operating costs per mile of vehicles of class/vintage c_n;
X is a vector of other explanatory variables (household and vehicle characteristics) that affect the utility the household can obtain from n vehicles of class/vintage c_n.

Given n and c_n, the number of miles the household will travel in its ith vehicle, g^i, can be derived by Roy's identity:

$$VMT^i_{nc_n} = - \frac{\partial V_{nc_n} / \partial P^i_{nc_n}}{\partial V_{nc_n} / \partial Y} = - g^i(Y, P_{nc_n}, X_{nc_n}) \qquad i = 1,...,n$$

For households owning only one vehicle, the utility derived from driving that vehicle depends upon household characteristics, vehicle characteristics and the vehicle's operating cost. For households owning more than one vehicle, the utility derived from driving its ith vehicle depends upon household characteristics and the characteristics and operating cost of each of its vehicles.

Vehicle characteristics, including make, model and year, are determined at the time of purchase.[8] Operating cost is affected by vehicle-specific fixed factors which affect the rate of fuel consumption, such as vehicle weight, engine type and size, and the emissions control system, as well as fuel price and vehicle maintenance. With maintenance held constant in the short run, operating cost varies directly with the price of fuel. For a vehicle of class/vintage c_n:

$$Pc_n = \$/gallon * gallons/VMT$$

where:

$\$/gallon$ = fuel price paid at the pump;
$gallons/VMT$ = the inverse of the vehicle's fuel efficiency = $1/MPG$.

Given a constant fuel price, a vehicle's cost per mile varies inversely with MPG. Operating costs can vary across vehicles achieving the same MPG if the vehicles use different grades or types of fuel. It is assumed, however, that all vehicles of a given class/vintage choose the same grade and type of fuel for optimal performance and, consequently, face the same cost per gallon.

A REVIEW OF VMT MODELS: TRAIN'S MODEL OF HOUSEHOLD VMT

Train (1986) estimated a model in which vehicle ownership and vehicle use are chosen jointly by each household. Separate submodels were estimated for the number of vehicles to own; the type (class and year) of vehicle or vehicles to own if the household owns any vehicle(s); and the annual VMT of each vehicle owned. Separate regression models were estimated for one- and two-vehicle households. The one-vehicle household's indirect utility function is:

$$V_{1c_1 m_{c_1}} = (1/1-\alpha_1))Y^{1-\alpha_1} + (1/\beta_1)\exp\{\delta_1 - \beta_1 p_{c_1 m_{c_1}}\} + \theta_{1c_1 m_{c_1}}$$

The two-vehicle household's indirect utility function is:

$$V_{2c_2m_{c_2}} = (1/1-\alpha_2))Y^{1-\alpha_2} + (1/\beta_2)\exp\{\delta_2 - \beta_2 p^1_{2c_2m_{c_1}}\}$$
$$- (1/\beta_2)\exp\{\delta_2 + \beta_2 p^2_{2c_2m_{c_2}}\} + \theta_{2c_2m_{c_2}}$$

where:

p is the cost in cents per mile of driving a vehicle of the class/vintage c_n and make/model m_{cn};

α_1, α_2, β_1, and β_2 are parameters;

δ_1 is a weighted sum of both observed and unobserved characteristics of the household, with the weights being parameters;

δ_2 is another weighted sum (not necessarily the same as δ_1) of household characteristics, with the weights being parameters;

$\theta_{1c_1m_{c_1}}$ is a weighted sum of both observed and unobserved household characteristics and characteristics of a vehicle of class/vintage c_1 and make/model m_{c1};

$\theta_{2c_2m_{c_2}}$ is a weighted sum of both observed and unobserved household characteristics and characteristics of the pair of vehicles of class/vintage c_2 and make/model m_{c2} where c_2 is a vector denoting the class/vintage of each of the two vehicles and m_{c2} is a vector denoting the make/model of each of the two vehicles.

Using Roy's identity, Train derived equations for VMT in the one-vehicle household and VMT for each vehicle in the two-vehicle household:

$$\text{One-vehicle household VMT} = \alpha_1 - \beta_1 p_{1c_1m_{c_1}} + \delta_1$$

For two-vehicle households, the VMT in each vehicle is:

$$VMT^i = \alpha_2 lnY - \beta_2 p^i_{2c_2m_{c_2}} + \delta_2, \quad \text{for } i=1,2$$

The functional form to estimate VMT in the one-vehicle household was specified as:

log(VMT) = β_0+β_1*log HH income+β_2*operating cost (cents per mile)+β_3*log HH size+β_4*number of workers in HH+β_5*number of transit trips per capita in area+β_6*urban dummy (>1 million population)+β_7*urban dummy (<1 million population)+ β_8*northeast US dummy+β_9*midwest US dummy+β_{10}*southern US dummy

The *i*th vehicle's VMT in a two-vehicle household was specified as:

$log(VMT_i) = \beta_0 + \beta_1 * log\ HH\ income + \beta_2 * operating\ cost_i + \beta_3 * newer$
vehicle dummy $+ \beta_4 * log\ HH\ size + \beta_5 * number\ of\ workers\ in\ HH +$
$\beta_6 * number\ of\ transit\ trips\ per\ capita\ in\ area + \beta_7 * urban\ dummy\ (>1$
million population) $+ \beta_8 * urban\ dummy\ (<1\ million\ population) +$
$\beta_9 * northeast\ US\ dummy + \beta_{10} * midwest\ US\ dummy + \beta_{11} * southern\ US$
dummy

Train argued that the household choice of vehicle implied the choice of operating cost, making operating cost endogenous. To avoid an endogeneity bias, Train used an instrument for operating cost based on: gasoline price; household income; household size; type of housing unit; population of household's area of residence; the number of transit trips in the area of residence; the number of adults; number of adolescents; number of workers; age of head of household; education level of head of household; gender of head of household; and distance to work (Train 1986, pp. 163–6).

The only vehicle characteristics used in Train's VMT models are the operating costs and the relative vehicle ages.

THE WALLS ET AL. MODEL OF HOUSEHOLD VMT

Walls et al. (1993) developed a series of household VMT models using the following general specification:

$$ln\ VMT^i = f(ln(Pnc_n), lnY, lnY * lnP^i, Xnc_n)$$

where:

VMT^i is the demand for travel in the ith vehicle;
Pnc_n is the vector of operating cost per mile for each vehicle of class/vintage c_n;
Y is household income;
P^i is the operating cost per mile of the ith vehicle;
Xnc_n is a vector of household demographic characteristics.

Like Train, Walls et al. fit separate models for one-vehicle and multi-vehicle households. One-vehicle household VMT was specified as:

$log(VMT) = \beta_0 + \beta_1 * northeast\ US\ dummy + \beta_2 * north\ central\ US\ dummy +$
$\beta_3 * southern\ US\ dummy + \beta_4 * western\ US\ dummy + \beta_5 * ln(own\ P) * NE$
dummy $+ \beta_6 * ln(own\ P) * NC\ dummy + \beta_7 * ln(own\ P) * S\ dummy + \beta_8 * ln(own$
$P) * W\ dummy + \beta_9 * ln(income) + \beta_{10} * ln(income) * ln(own\ P) + \beta_{11} * children$
in HH dummy $+ \beta_{12} * retired,\ no\ children\ dummy + \beta_{13} * urban\ dummy +$

β_{14}*suburban dummy+β_{15}*number of drivers in HH+β_{16}*other vehicle in HH dummy+β_{17}*public transit availability dummy+β_{18}*one main driver for vehicle dummy+β_{19}*vehicle owned (not leased) dummy+ β_{20}*highest education level in HH+β_{21}*vehicle bought new within last year dummy

The model of the *i*th vehicle's VMT in the two-vehicle household included all the independent variables used in the one-vehicle household specification plus a dummy indicating whether the vehicle is the newer one and four interacted cross-price terms: *ln(cross P)*NE*; *ln(cross P)*NC*; *ln(cross P)*S* and *ln(cross P)*W*. The cross-price term measured the operating cost of the other household vehicle.

For households with three or more vehicles, the cross-price term was the average operating cost of all other vehicles in the household. The three-or-more-vehicle model also included the number of vehicles in the household as an independent variable.

VEHICLE CHARACTERISTICS AND VMT: AN OVERVIEW

Train and Walls et al. posit that household VMT depends, in part, on a vector of vehicle characteristics. Their models include vehicle operating costs and, in the case of multi-vehicle households, the relative newness of the vehicles. However, an analysis of the 1991 national *RTECS* data shows that vehicle ages and operating costs (determined by fuel efficiency) do not consistently explain household vehicle usage choice in the multi-vehicle household.

Table 3.4 shows that the majority of two-vehicle *RTECS* households drove more miles in their newer vehicle and more miles in their more fuel-efficient vehicle but only 45.2 per cent of the two-vehicle households drove the majority of their annual miles in a vehicle that was both newer and more fuel-efficient. Households owning two vehicles of the same age were indifferent between driving more miles in their higher or lower MPG vehicle.

Table 3.5 shows that, among three-vehicle households with different VMT, MPG and vehicle ages, only 25.6 per cent drove the most miles in their newest and most fuel-efficient vehicle, while 46 per cent drove the most miles in their newest vehicle, and 45.8 per cent drove the most miles in their most fuel-efficient vehicle.

There are several important factors underlying this observed behaviour:

1. Households may not perceive or respond to small differences in MPG when choosing a vehicle for a particular trip.[9] Similarly, vehicles that are almost

Table 3.4 Vehicle chosen for the majority of use in two-vehicle households

	Lower MPG vehicle	Higher MPG vehicle	Same MPG both vehicles	Same VMT different MPGs	Total
Older vehicle	180 (16.7%)	152 (14.1%)			322 (30.8%)
Newer vehicle	160 (14.9%)	487 (45.2%)	5 (0.5%)		652 (60.5%)
Same age vehicles	42 (3.9%)	41 (3.8%)		2 (0.2%)	85 (7.9%)
Same VMT, different ages				8 (0.7%)	8 (0.7%)
Total	382 (35.5%)	680 (63.1%)	5 (0.5%)	10 (0.9%)	1077 (100%)

Table 3.5 Vehicle with the highest VMT in three-vehicle households in which all vehicles are of different ages, VMT and MPG

	Lowest MPG vehicle	Middle MPG vehicle	Highest MPG vehicle	Total
Oldest vehicle	38 (8.8%)	26 (6.0%)	25 (5.8%)	89 (20.7%)
Middle vehicle	28 (6.5%)	53 (12.3%)	63 (14.7%)	144 (33.5%)
Newest vehicle	24 (5.6%)	63 (14.7%)	110 (25.6%)	197 (45.8%)
Total	90 (21.0%)	142 (33.0%)	198 (46.0%)	430 (100%)

the same age may not be perceived as newer or older.

2.   The newer vehicle may have been owned for less than a full year. Although *RTECS* results are presented as annual or annualized data, it is possible that some late model vehicles' VMTs are not annualized correctly.

3.   In many households, each vehicle is typically driven frequently or exclusively by only one household member. This is especially true for the trip to work. A vehicle's VMT may be primarily a function of a household member's distance to work or school.

4.   Newer vehicles are not necessarily more fuel-efficient. In particular,

minivans, pick-up trucks and sport-utility vehicles have relatively low fuel efficiency but have become popular as household vehicles. Some cars produced in the late 1970s are more fuel-efficient than late model vehicles.

5. Vehicle characteristics other than age and fuel efficiency may play a strong role in determining which vehicle to use for travel under certain conditions. A vehicle that is satisfactory for single-occupant commuting may not be appropriate for multi-person or long-distance travel. Similarly, road or weather conditions may encourage the use of a less fuel-efficient four-wheel drive vehicle with better traction.

6. Americans appear to have a taste for large, powerful vehicles, despite their low fuel efficiency. As Table 3.6 shows, light trucks, which include pick-ups, vans, minivans and sport-utility vehicles, accounted for approximately one-third of all new vehicle sales in 1991 and 1992 (American Automobile Manufacturers Association (AAMA) 1993, p. 53). The federal fuel efficiency standard for these vehicles is 26.5 per cent below that for passenger cars (20.2 MPG for light trucks vs 27.5 MPG for cars).

Table 3.6 Percentage of new vehicles sold in the US by vehicle type

	1991	1992
Small except specialty	19.1%	18.5%
Large + luxury	14.7%	14.4%
Sport-utility	7.4%	8.8%
Vans, minivans	9.6%	10.5%
Pick-up trucks	16.5%	16.7%
Total light trucks	33.5%	36.0%

Note: Light trucks = sport-utility + vans + minivans + pick-up trucks

Source: AAMA (1993).

The 1991 *RTECS* data strongly suggest that vehicle characteristics other than age and fuel efficiency affect household vehicle usage. Models of vehicle ownership choice, such as Train (1986), have shown that vehicle characteristics such as capacity, horsepower and shoulder space are important determinants of the type of vehicle a household will choose to own. *RTECS* data suggest that vehicle characteristics are also important determinants of vehicle use. With households free to substitute among their vehicles, and with vehicle age and fuel efficiency explaining only part of the substitution effects, it is likely that both the Train and Walls et al.'s VMT models overestimate the effect of operating costs on vehicle usage by omitting physical characteristics of the vehicles themselves.

Vehicle characteristics have been introduced with some success in other household vehicle usage models. Hensher (1985) included two vehicle-type dummies in his vehicle usage models for two-vehicle and three-vehicle households in Australia: panel van or utility vehicle, and light commercial or camper van. He found the panel van or utility vehicle dummy was significant in the three-vehicle model only and the other dummy was insignificant in both models.

A recent model developed by Golub et al. (1994) estimates VMT by vehicle type, using household, principal driver and vehicle characteristics. Vehicle characteristics include vehicle age, operating cost per mile and 12 dummies to describe vehicle type. The Golub et al. model was designed to analyse the allocation of drivers in the household to specific vehicles. The dummies for subcompact, sports car, compact pick-up truck, minivan, standard van, small sport-utility and standard sport-utility vehicles were all significant in their reduced-form model.

AN EXPANDED MODEL OF VMT USING VEHICLE CHARACTERISTICS

The elasticity of operating cost per mile is the key measure of the effect of the emissions tax on travel. The operating cost per mile will become the sum of the fuel cost per mile plus the emissions fee per mile:

$$Pnc_n = \$/gallon * gallons/mi + \sum_{j=1}^{3} \$/gram^j * grams^j/mi$$

where:

$\$/gram^j$ = tax rate per gram of the jth pollutant (j=HC, CO, NO$_x$);
$grams^j/mi$ = emissions rate per mile of the jth pollutant.

In order to develop sound estimates of the own- and cross-price elasticities, separate ordinary least squares regression models are specified for households owning one, two and three or more vehicles, following the general approach of Walls et al. Three classes of independent variable are used: household characteristics; physical characteristics of each household vehicle; and a dummy indicating whether the vehicle is used to commute to work. The models are estimated using the 1991 *RTECS* household data.

Household characteristics affect the decision to travel by personal vehicle, regardless of the number of vehicles owned. These characteristics include the household size, the age of the head of household and household income.

Physical vehicle characteristics used in the models include operating cost per mile, age and vehicle type. Operating cost is calculated from the vehicle's fuel efficiency (MPG) and the average cost of gasoline. Fuel efficiency is a function of imbedded vehicle characteristics such as the type of fuel system and pollution control equipment, as well as vehicle maintenance, driving speed and roadway conditions. The individual driver takes fuel price and roadway conditions as given and chooses maintenance and speed. The driver may also choose roadway conditions through his or her choice of route or time of day to travel. The VMT models assume a vehicle's fuel efficiency is constant and equal to the MPG reported in the *RTECS* data. Fuel cost is the cost per gallon reported by *RTECS*.

Vehicle type is based on the *RTECS* categories of passenger car, station wagon, minivan, full-sized van, pick-up truck and sport-utility vehicle. There is no further distinction made regarding types of passenger car.

THE ONE-VEHICLE HOUSEHOLD VMT MODEL

The following independent variables are used to estimate the log of VMT for one-vehicle households:

1. *Geographic dummies*:
   MIDWEST, SOUTH, MTN and *PACIFIC* were defined based on census codes. The omitted region was the Northeast. The Mountain and Pacific regions are sub-divisions of the West region.
2. *Demographic variables*:
   HHAGE is the age of the head of the household;
   HHSIZE is the number of household members;
   COMMUT is a dummy indicating that the vehicle is used for commuting to work. This captures an implied characteristic of household member employment and also implies a minimum level of vehicle usage.
   LINCOME is the log of household income. Income was classified in ranges in the *RTECS* dataset. Household income has been estimated as the mean of the indicated range. As with many income variables, it is likely that in-kind transfers and other income supplements were ignored by households reporting very low incomes.
3. *Vehicle characteristic variables*:
   LCOSTMI is the log of the vehicle operating cost per mile;
   MODELYR is the vehicle vintage or model year;
   SWAGON is a dummy indicating the vehicle is a station wagon;
   VAN is a dummy indicating the vehicle is a full-sized van;
   MINIVAN is a dummy indicating the vehicle is a minivan;
   PICKUP is a dummy indicating the vehicle is a pick-up truck;

SPORTUT is a dummy indicating the vehicle is a sport-utility type.
The omitted vehicle type is the passenger car, which ranges from
subcompact to luxury, from sports car to sedan. As Table 3.7 shows,
nearly 80 per cent of vehicles owned by one-vehicle households are
passenger cars.

Table 3.7 Types of vehicle owned by one-vehicle households

Type of vehicle	Number of households	% of households
Passenger car	640	79.4
Station wagon	45	5.6
Full-sized van	10	1.2
Minivan	13	1.6
Pick-up truck	79	9.8
Sport-utility vehicle	19	2.4

Source: Derived by the author from 1991 *RTECS* data.

The results of the OLS estimate of household VMT in the one-vehicle
household are given in Table 3.8. These results indicate that VMT declines as
the age of the head of household and vehicle operating costs increase, and
increases with income and the use of the vehicle for commuting. Non-passenger
cars are driven more, on average, than passenger cars, although this effect is
statistically significant only for sport-utility vehicles. The vehicle age was
omitted from this specification because of its insignificant effect on VMT in the
one-vehicle household.

The price elasticity of VMT demand is higher than that estimated by several
other studies, as shown in Table 3.9. However, note that all of the studies listed
in Table 3.9 show VMT to be price-inelastic for one-vehicle households.

THE TWO-VEHICLE HOUSEHOLD VMT MODEL

The two-vehicle household can substitute among its vehicles when it decides to
make a trip. Operating costs and the physical characteristics of both vehicles
can influence the household's demand for travel in each of its *i* vehicles.
Household demographics, as described above, also affect the two-vehicle
household's VMT decisions.

VMT of each vehicle is estimated as a function of household characteristics,
geographic location, characteristics of the vehicle itself and characteristics of the

Table 3.8 One-vehicle household OLS model

Dependent Variable: *LVMT*

(a) Analysis of variance

Source	DF	Sum of squares	Mean square	F value	Prob > F
Model	14	137.74281	9.83877	15.495	0.0001
Error	791	502.25455	0.63496		
C total	805	639.99736			
Root MSE	0.79684	R-square	0.2152		
Dep mean	8.90258	Adj R-sq	0.2013		
CV	8.95072				

(b) Parameter estimates

Variable	DF	Parameter estimate	Standard error	T for H0	Prob > | T |
INTERCEP	1	9.697766	0.436833	22.200	0.0001
MIDWEST	1	0.026491	0.081597	0.325	0.7455
SOUTH	1	−0.056960	0.077975	−0.730	0.4653
MTN	1	0.014836	0.120695	0.123	0.9022
PACIFIC	1	−0.132874	0.095783	−1.387	0.1658
HHSIZE	1	0.001980	0.021923	0.090	0.9281
HHAGE	1	−0.005579	0.001793	−3.112	0.0019
LINCOME	1	0.091671	0.036928	2.482	0.0133
LCOST	1	−0.851894	0.096108	−8.864	0.0001
SWAGON	1	0.084224	0.123606	0.681	0.4958
VAN	1	0.372653	0.257786	1.446	0.1487
MINIVAN	1	0.293543	0.225280	1.303	0.1929
PICKUP	1	0.185484	0.097628	1.900	0.0578
SPORTUT	1	0.509627	0.187940	2.712	0.0068
COMMUT	1	0.244962	0.069952	3.502	0.0005

Table 3.9 Price elasticity estimates for one-vehicle households

Study	Elasticity
This study	−0.852
Archibald and Gillingham (1980)	−0.430
Hensher (1985)	−0.452
Hensher et al. (1992)	−0.222
Mannering and Winston (1985)	−0.228
Walls et al. (1993)	−0.288

other vehicle. The same geographic and demographic variables used for the one-vehicle household model are also used in the two-vehicle model: *MIDWEST*, *SOUTH*, *MTN*, *PACIFIC*, *HHSIZE*, *HHAGE*, *LINCOME* and *COMMUT1*. *COMMUT1* indicates that the vehicle of interest, vehicle 1, is used for commuting to work. The use of the other vehicle for commuting is expected to have no effect on the decision to use vehicle 1 and is not used in this specification.

Vehicle characteristics:
LCOSTMI1, LCOSTMI2: operating cost per mile for vehicle 1 and vehicle 2;
SWAGON1, SWAGON2: dummy indicating vehicle 1 or 2 is a station wagon;
VAN1, VAN2: dummy indicating vehicle 1 or 2 is a van;
MINIVAN1, MINIVAN2: dummy indicating vehicle 1 or 2 is a minivan;
PICKUP1, PICKUP2: dummy indicating vehicle 1 or 2 is a pick-up truck;
SPORTUT1, SPORTUT2: dummy indicating vehicle 1 or 2 is a sport-utility.

As Table 3.10 shows, 89.2 per cent of the 1077 sampled households owning two vehicles owned at least one passenger car and 32.3 per cent owned at least one pick-up truck. Only 42.0 per cent of the two-vehicle households owned passenger cars for both of their vehicles, while 25.1 per cent owned one passenger car and one pick-up truck.

The estimate of VMT for each vehicle in the two-vehicle household is shown in Table 3.11. As in the one-vehicle household, VMT in the *i*th vehicle increases with household income, household size, the newness of the vehicle and the use of the vehicle for commuting to work, and declines with increases in operating cost and the age of the head of household. VMT also increases if the *i*th vehicle is something other than a car, although the effect is significant only for pick-up trucks or sport-utility vehicles. Vehicle type for the other household vehicle does not have a significant effect on vehicle *i*'s VMT.

Table 3.10 Types of vehicle owned by two-vehicle households

Vehicle combination	Number of households	% of households
Both cars	452	42.0
Car, wagon	78	7.2
Car, van	34	3.2
Car, minivan	59	5.5
Car, pick-up	267	24.8
Car, sport-utility	71	6.6
Both wagons	6	0.6
Wagon, van	2	0.2
Wagon, minivan	4	0.4
Wagon, pick-up	26	2.4
Wagon, sport-utility	10	0.9
Both vans	3	0.3
Van, minivan	0	0.0
Van, pick-up	6	0.6
Van, sport-utility	2	0.2
Both minivans	2	0.2
Minivan, pick-up	17	1.6
Minivan, sport-utility	3	0.3
Both pick-ups	11	1.0
Pick-up, sport-utility	21	1.9
Both sport-utility	3	0.3

Source: Derived by the author from 1991 *RTECS* data.

The own-price elasticity shown in Table 3.11 for the two-vehicle household, −0.9155, is significant and more elastic than that derived for the one-vehicle household (−0.8519) in Table 3.8. The cross-price elasticity is positive, as expected, at 0.1378, but is not quite significant. This result, combined with the lack of significance for the other household vehicle type, suggests that the two-vehicle household is less likely to substitute between vehicles than previously expected. It may be that these households typically have a dedicated driver for each vehicle and the vehicle's VMT is less a function of household needs than it is of individual driver needs.

Table 3.11 Two-vehicle household OLS model

Dependent Variable: *LVMT1*

(a) Analysis of variance

Source	DF	Sum of squares	Mean square	F value	Prob > F
Model	21	603.57511	28.74167	21.386	0.0001
Error	2132	2865.27284	1.34394		
C total	2153	3468.84794			

Root MSE	1.15928	R-square	0.1740	
Dep mean	8.86725	Adj R-sq	0.1659	
CV	13.07376			

(b) Parameter estimates

Variable	DF	Parameter estimate	Standard error	T for H0	Prob > | T |
INTERCEP	1	5.555391	0.690774	8.042	0.0001
MIDWEST	1	0.018518	0.076382	0.242	0.8085
SOUTH	1	0.011147	0.077351	0.144	0.8854
MTN	1	−0.063122	0.107787	−0.586	0.5582
PACIFIC	1	−0.230753	0.085383	−2.703	0.0069
HHSIZE	1	0.050690	0.019892	2.548	0.0109
HHAGE	1	−0.005554	0.001821	−3.051	0.0023
MODELYR1	1	0.036649	0.005823	6.294	0.0001
LCOST1	1	−0.915593	0.104358	−8.774	0.0001
LCOST2	1	0.137830	0.086886	1.586	0.1128
LINCOME	1	0.154907	0.036774	4.212	0.0001
SWAGON1	1	0.139688	0.107056	1.305	0.1921
VAN1	1	0.170514	0.173174	0.985	0.3249
MINIVAN1	1	0.218030	0.133003	1.639	0.1013
PICKUP1	1	0.311753	0.073845	4.222	0.0001
SPORTUT1	1	0.303679	0.119667	2.538	0.0012
SWAGON2	1	−0.072110	0.107046	−0.674	0.5006
VAN2	1	0.017526	0.173105	0.101	0.9194
MINIVAN2	1	−0.114888	0.131150	−0.876	0.3811
PICKUP2	1	0.090772	0.073580	1.247	0.2124
SPORTUT2	1	0.079712	0.117640	0.678	0.4981
COMMUT1	1	0.163984	0.056552	2.900	0.0038

THE THREE-OR-MORE VEHICLE HOUSEHOLD VMT MODEL

Like the two-vehicle household, the three-or-more-vehicle household can substitute among its vehicles. Operating costs, vehicle characteristics and type, and household demographics, as described above, also affect VMT decisions.

The three-or-more-vehicle household VMT model is similar to the two-vehicle household VMT model, with the following modifications:

1. The average operating cost for all other household vehicles is used to estimate the cross-price elasticity, consistent with the approach used by Walls et al.
2. Each household vehicle in turn is considered vehicle number 1 and the log of its VMT is the dependent variable. Vehicle-type dummies are defined for all the other household vehicles and entered into the regression as *SWAGON2, VAN2, MINIVAN2, PICKUP2* and *SPORTUT2*. Multiple occurrences of these vehicle types are ignored (that is, if the household has one or more pick-up trucks among its other vehicles, *PICKUP2*=1).
3. A new variable, *NUMVEH*, is introduced where *NUMVEH* equals the number of vehicles owned by the household ($3 \leq NUMVEH \leq 10$).

Table 3.12 shows the results of the three-or-more-vehicle household VMT model. Like the two-vehicle household model, VMT in the ith vehicle increases with household size, income, vehicle newness, the use of the vehicle for commuting and an increase in the average operating cost of the household's other vehicles, and decreases with the age of the head of household and the vehicle's own operating costs. Vehicles other than cars have higher VMT, with the effect being significant for pick-ups, sport-utility vehicles and vans. VMT in the ith vehicle is inversely related to the number of household vehicles.

DISCUSSION OF VMT MODELS AND ELASTICITIES

Given the greater substitution possibilities, it is not surprising that the cross-price elasticity in the three-or-more-vehicle household (0.3011) is higher than that in the two-vehicle household (0.1378). The own-price elasticity for the three-or-more-vehicle household (−0.9410) is virtually identical to that of the two-vehicle household (−0.9156), perhaps indicating the presence of a primary driver for each of the *NUMVEH* vehicles in the multi-vehicle household.

Table 3.9 showed that the own-price elasticity estimated for one-vehicle households is considerably larger than that estimated by most other studies. Several of the other studies estimated models for two-vehicle households, but

Table 3.12 Three-or-more-vehicle household OLS model

Dependent Variable: *LVMT1*

(a) Analysis of variance

Source	DF	Sum of squares	Mean square	F value	Prob > F
Model	22	571.33556	25.96980	29.009	0.0001
Error	3080	2757.33535	0.89524		
C total	3102	3328.67091			

Root MSE	0.94617	R-square	0.1716
Dep mean	8.68092	Adj R-sq	0.1657
CV	10.89943		

(b) Parameter estimates

Variable	DF	Parameter estimate	Standard error	T for H0	Prob > | T |
INTERCEP	1	7.562497	0.463177	16.327	0.0001
MIDWEST	1	−0.046158	0.050586	−0.912	0.3616
SOUTH	1	0.036028	0.052024	0.693	0.4887
MTN	1	0.116373	0.070365	1.654	0.0983
PACIFIC	1	0.044115	0.061342	0.719	0.4721
HHSIZE	1	0.041670	0.013176	3.163	0.0016
HHAGE	1	−0.000880	0.001337	−0.658	0.5108
MODELYR1	1	0.014947	0.003272	4.568	0.0001
LCOST1	1	−0.940961	0.065333	−14.402	0.0001
LCOST2	1	0.301147	0.076512	3.936	0.0001
LINCOME	1	0.106074	0.028477	3.725	0.0002
SWAGON1	1	0.001956	0.081450	0.024	0.9808
VAN1	1	0.531203	0.103082	5.153	0.0001
MINIVAN1	1	0.109070	0.097851	1.115	0.2651
PICKUP1	1	0.271303	0.045444	5.970	0.0001
SPORTUT1	1	0.246951	0.079958	3.089	0.0020
SWAGON2	1	−0.037713	0.053565	−0.704	0.4814
VAN2	1	0.084445	0.064666	1.306	0.1917
MINIVAN2	1	−0.186414	0.064472	−2.891	0.0039
PICKUP2	1	−0.013895	0.038060	−0.365	0.7151
SPORTUT2	1	−0.072918	0.052104	−1.399	0.1618
COMMUT1	1	0.209541	0.037605	5.572	0.0001
NUMVEH	1	−0.094227	0.020824	−4.525	0.0001

most did not estimate separate models for three-or-more-vehicle households. Where all three models were estimated, as in the Walls et al. study, the elasticities from the two- and three-or-more-vehicle models are not remarkably different from those estimated in the present study, as shown in Table 3.13.

Table 3.13 Comparison of elasticities from two studies

	Walls et al.	This study
One-vehicle household:		
own-price elasticity	−0.13	−0.85
Two-vehicle household:		
own-price elasticity	−0.52	−0.92
cross-price elasticity	0.20	0.14
Three-or-more-vehicle household:		
own-price elasticity	−0.92	−0.94
cross-price elasticity	−0.04	0.30

While it is easy to pin the differences on the choice of datasets and model specifications, it is important to consider some of the other factors that may be responsible for these elasticity differences, especially in the one-vehicle models.

There is a large literature on the demand for gasoline which has produced a wide range of estimates of the elasticity of gasoline demand with respect to price. Dahl's (1986) survey showed the price elasticity of gasoline demand ranging from −2.42 to +0.37, depending on the type of model and frequency of data (Dahl, 1986, pp. 68–71). The long-run gasoline demand price elasticity, derived from cross-sectional analyses, averaged −1.02.

The price elasticity of gasoline demand can be rewritten as the price elasticity of VMT minus the price elasticity of MPG. Because the latter term should be positive[10] (Dahl's survey showed a range of 0.06 to 1.43), VMT elasticity should be less than that of gasoline. The elasticity of VMT with respect to gasoline price ranged from −2.25 to +1.9 in the models Dahl surveyed. This is a substantial range.

A survey by Drollas (1984) showed VMT elasticities ranging from −0.36 to −0.52. He estimated the long-run gasoline demand elasticity to be around −0.80 in the United States (Drollas, 1984, p. 73).

Dahl and Sterner (1991) reported a wide range of gasoline price elasticities, with cross-sectional models being the most elastic (averaging −1.01). Goodwin (1992) also concluded that cross-sectional models produce more elastic price demand for gasoline or VMT than time series or cross-section time series models.

What accounts for the generally high price elasticities of VMT or gasoline demand in the cross-sectional models? There are several factors to consider. Cross-sectional analysis represents a long-run equilibrium, implying that all adjustments to the vehicle stock have already been made. Models which ignore vehicle stock variables, as many of the models in the above-mentioned surveys do, may overestimate the price elasticity because of an omitted variable bias.

A second problem, which this present study also encounters, is the lack of an explicit model of vehicle choice. By taking the vehicle stock as given, the simultaneous nature of the vehicle choice/vehicle usage decision is ignored. Train (1986) approached the problem by modelling the number of vehicles for a household to own, the type or types of vehicle(s) to own and the VMT for each vehicle, but most other studies have not modelled the vehicle choice.

Vehicle choice models endogenize both the vehicle stock and the operating cost. When the vehicle is chosen, its characteristics, including the type of fuel and the fuel efficiency, are also chosen. Given a fuel price that is exogenous to the vehicle owner, the fuel cost per mile is effectively chosen with the vehicle. This same fuel cost, in turn, is used to determine the demand for gasoline or VMT in the vehicle usage models. It is possible that the fuel price elasticity is overstated because the fuel price per mile has not been determined endogenously in the vehicle usage models.

Another potential source of estimation error may come from the definition of the operating cost term itself. All the models reviewed consider the fuel cost per mile to be the relevant price. This ignores other operating costs, including oil, tyres and other wear and tear, the cost of travel time and ownership costs, including licensing, registration, insurance, financing and depreciation. If these other costs are added to the fuel cost, the importance of the fuel cost in the travel decision would be diminished. These additional costs belong in the vehicle choice model which, again, is absent from this analysis. This may lead to the operating cost elasticity being overstated.

The inclusion of vehicle characteristics in the VMT models presented in this study attempts to develop a stronger link between the vehicle choice and usage decisions, without explicitly modelling the vehicle choice decision. If this specification is correct, the relatively high price elasticity found in the one-vehicle household model may be a better reflection of a long-run equilibrium than the elasticity estimates of other models which do not consider the vehicle characteristics. Further study is clearly indicated in this area.

The broad range of possible elasticities is troubling from a policy perspective because this indicates a wide range of potential response to an emissions tax. If the elasticities developed in this study are correct, they imply a very strong response to any increase in operating cost, regardless of the number of vehicles in the household. If other results are correct, they indicate a smaller reduction in VMT and, consequently, emissions from the same tax.

To handle this uncertainty, a VMT and emissions reduction simulation model

(TIERS) has been developed to test the sensitivity of emissions reduction to the VMT elasticities. Although a more inelastic VMT assumption produces smaller reductions in VMT and emissions, TIERS shows that significant reductions would be expected to occur even under the most inelastic scenario. The TIERS model, and its resulting estimates of VMT and emissions reductions under different elasticity assumptions, is discussed at length in Chapter 4.

NOTES

1. The four-step models described here are adapted from Stopher and Meyburg (1975). Similar discussions can be found in any standard text on travel demand analysis and planning.
2. Stopher and Meyburg (1975, p. 140) trace the use of the gravity model in trip distribution to William J. Reilly (1931, p. 9) who posited that, 'two cities attract retail trade from any intermediate town or city ... approximately in direct proportion to the population of the two cities and in inverse proportion to the square of the distance of these cities from the intermediate town'.
3. The *NPTS* was conducted by the US Bureau of the Census under contract to the US Department of Transportation in 1969, 1977 and 1983 using nationally representative samples. The 1990 *NPTS* was conducted under contract with the Research Triangle Institute of North Carolina. The most recent survey was telephone-based. Despite changing samples and methodology, the preference for personal vehicles has been relatively constant across the four surveys. Mention of the 1990 *NPTS* in this study refers to the summary prepared by Pisarski (1992).
4. The emissions tax targets VMT and maintenance or vehicle replacement to reduce emissions per mile. Given the strong preference for personal vehicle travel, this study has chosen to ignore the effect of induced mode substitution on emissions reduction. Studies such as Cambridge Systematics, Inc. (1993) and Cameron (1991) indicate that VMT pricing would reduce trip generation and increase public transit's mode share of trips, but personal vehicle travel would continue to be the dominant mode.
5. See, for example, Stedman (1989), Lawson (1993) and Cadle et al. (1993).
6. Vehicle choice is not exogenous but it is held constant in the short run. The household chose its vehicle(s) with no anticipation of an emissions tax.
7. Vehicle class is based on vehicle characteristics and is used interchangeably with vehicle type (for example, van, sports car) in this discussion. Vintage refers to model year and is a surrogate for vehicle age at a point in time.
8. Some characteristics, such as the audio system, can be modified subsequent to vehicle purchase. However, the major differences in vehicle type (such as sports car, van or station wagon) and age are fixed.
9. Malecki's (1978) survey showed that perceived and actual operating costs for motor vehicles often diverged, with motorists overestimating the fuel costs of their journey to work for short distance commutes.
10. When the price of gasoline increases, motorists would be expected to increase maintenance or replace existing vehicles with more fuel-efficient ones. This produces a positive elasticity of MPG with respect to gasoline price.

4. The TIERS Model

The TIERS model assesses the expected impact of the emissions tax on VMT and calculates the resulting emissions reduction. The elasticity of VMT with respect to the cost per mile of travel is a key determinant of the household's response to the tax. The own-price elasticity for one-vehicle households and the own- and cross-price elasticities for two- and multi-vehicle households are used to determine the response of each type of household separately. The travel responses, and resulting changes in emissions, are aggregated across all households using demographic weights provided by the California Department of Transportation (CalTrans) 1991 household travel survey.

By default, TIERS uses the elasticities developed in Chapter 3 and the tax rates developed in Chapter 2, but TIERS can be run with any set of tax rates (including zero for any or all pollutants) and any set of own-price and cross-price elasticities. TIERS also simulates the effect of a change in the gasoline tax, either alone or in conjunction with the emissions tax.

TIERS' flexibility makes it especially well suited to analysing the sensitivity of emissions reduction to different assumptions of price elasticity and tax rates. This can be useful both for the analyst and the policy-maker. It must be emphasized, however, that the TIERS results presented here are specific to the South Coast Air Basin in 1990.

TECHNICAL DESCRIPTION OF THE TIERS MODEL

TIERS works as follows:

1. Households in the CalTrans household travel survey in the four counties comprising the South Coast Air Basin are sorted into one-vehicle, two-vehicle and three-or-more-vehicle household groups, and each group is written to a different dataset.
2. Each household record is processed individually. Because the household records do not contain information on vehicle emissions, each vehicle is assigned a random emissions reading based on the vehicle model year.[1] The random draw is obtained from the USEPA I/M 240 data collected in

Hammond, Indiana which contains readings of HC, CO and NO_x for 10149 individual vehicles.

3. A fee per mile is calculated for the vehicle based on the emissions assigned:

$$Fee/mile = \sum_i fee/gram_i * emissions_i/mile \qquad i=\text{HC, CO, NO}_x$$

4. The change in each vehicle's VMT is calculated using the elasticities for one-, two- or multi-vehicle households, derived from the VMT models described in Chapter 3 or input by the user.
5. Aggregate VMT reduction is calculated as the weighted sum of the VMT reduced by each household for each of its k vehicles. The household weight is included in the CalTrans data record and is based on the household's demographic profile.
6. Aggregate emissions reduction is calculated for each pollutant based on the VMT reduction, the randomly assigned emissions/mile for each pollutant, and the household weight for each of its k vehicles in the n households:

$$Reduction_i = \sum_{j=1}^{n} \sum_{k=1}^{nveh} weight_j * VMT\ reduction_k * emissions_{ik}$$

where:

$i=\text{HC, CO, NO}_x$.

TIERS uses data from the following sources:

1. Household vehicle ownership (model year and vehicle type) plus household demographic weight from the CalTrans household travel survey of 1991.
2. Household demographics, vehicle ownership, vehicle usage (VMT), vehicle MPG and the fuel cost per mile from the US Department of Energy 1991 *RTECS* survey.
3. Emissions per mile of HC, CO and NO_x and vehicle model year from the USEPA's I/M 240 testing facility in Hammond, Indiana. Testing occurred between 1990 and 1992.
4. Average MPG for a vehicle of a given class and vintage (model year) from the California Energy Commission's (CEC's) PVMCOND model.
5. Average VMT for a vehicle of a given vintage in the South Coast region from the California Air Resources Board (CARB) 1990 household vehicle database.
6. Tax rates for HC, CO and NO_x are coded into the program, based on the lowest alternative control cost from stationary sources in the South Coast region, as discussed in Chapter 2. TIERS allows the user to override these

rates.

7. Price elasticities were derived outside TIERS using OLS models to estimate VMT for each vehicle in the one-vehicle, two-vehicle or three-or-more-vehicle households, as discussed in Chapter 3. TIERS allows the user to override these elasticities.

8. The user inputs an odd integer between 1 and 65535 as a seed value to start the random number generator. This follows from the FORTRAN-based random number generator of Hume and Holt (1985).

As Figure 4.1 shows, the distribution of passenger vehicles within the South Coast region in 1990, as reported by CARB, is very similar to the distribution of vehicles in the CalTrans household survey of 1991.

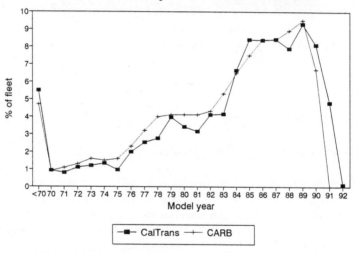

Age distribution of passenger vehicles: CalTrans survey and CARB database

Figure 4.1 Vehicle holdings in the South Coast Air Basin

The CalTrans households were chosen on the basis of their demographic profiles, not on the basis of the age of their vehicles. CARB data represent the total passenger car fleet in the South Coast region in 1990 by model year, and include estimates of total VMT and emissions per mile for each model year. CalTrans data do not include emissions or VMT. Given the similarity of the two distributions, and allowing for the difference in the time of the sampling (1990 vs 1991), it is reasonable to conclude that the CalTrans households are representative not only of the demographics of the region, but also of the vehicle holdings profile. For this reason, TIERS uses the model year data in the CARB database to analyse the expected changes in aggregate VMT and emissions induced by the emissions tax.

Appendix D contains the detailed programming flow of TIERS.

TIERS MODEL RESULTS

The TIERS model was designed for interactive use. The results presented here are representative of the VMT and emissions reduction that would be expected to result from the emissions tax. Results vary slightly with the choice of the random number used to seed the random number generator. Results vary substantially as the tax rates or elasticities are changed.

Table 4.1 contains a summary of travel and emissions impacts, given the tax rates discussed in Chapter 2 and the elasticities developed in Chapter 3. Each iteration produces results based on its own seed value for the random draw of emissions. Detailed results for one-vehicle, two-vehicle and multi-vehicle households' VMT and emissions changes by model year, as well as results aggregated across all households, are contained in the appendix to this chapter.

Table 4.2 shows the results of running the TIERS model with the same seed value (61255) and tax rates with the Walls et al. elasticities. Compared with the original results (Table 4.1), the VMT and emissions reductions in Table 4.2 are more modest. This is a direct result of the much more inelastic VMT demand for one-vehicle and two-vehicle households in the Walls et al. study. Tax revenues are also greater because the less elastic travel demand leaves VMT at a higher level in this scenario.

DISCUSSION OF TIERS RESULTS

The most important feature of the TIERS results is the sizeable emissions reduction projected, regardless of which elasticities are used in the simulation. This strongly suggests that the emissions tax has the potential to be a significant policy instrument in the South Coast, where substantial quantities of HCs (specifically ROG), CO and NO_x must be abated.

TIERS results in Table 4.1, using the author's elasticities, and Table 4.2, using Walls et al.'s elasticities, estimate that household vehicles' emissions would fall 53–76 per cent for HC, 48–72 per cent for CO and 39–62 per cent for NO_x. CARB data show that passenger cars produced 386 tons per day (TPD) of ROG, 2557 TPD of CO and 277 TPD of NO_x in 1990.[2]

Table 4.1 TIERS model: travel and emissions summary

Emissions tax rates ($/g): HC = $0.0165, CO = $0.00022, NO_x = $0.0088

One-vehicle household:	own-price elasticity	=	−0.85189
Two-vehicle household:	own-price elasticity	=	−0.91559
	cross-price elasticity	=	0.13783
Multi-vehicle household:	own-price elasticity	=	−0.94096
	cross-price elasticity	=	0.30115

Iteration	VMT base	% change	HC base	% change	CO base	% change	NO_x base	% change	Revenue ($mil)
1	112053	−41.0	95967	−75.6	1495330	−71.9	204613	−62.2	1160
2	112053	−40.5	95123	−75.5	1447596	−71.0	203202	−62.0	1156
3	112053	−40.8	96289	−75.7	1435247	−71.0	204679	−62.3	1156
4	112053	−40.8	98231	−76.7	1489415	−72.4	204870	−62.5	1144
5	112053	−40.9	96787	−76.1	1445413	−71.7	205984	−61.9	1163
6	112053	−41.1	98064	−76.3	1501377	−72.4	203911	−62.3	1150
7	112053	−41.4	97449	−76.4	1480044	−72.0	209683	−63.0	1153
8	112053	−40.9	96524	−76.1	1464844	−71.9	206972	−62.6	1153
9	112053	−40.8	95015	−75.5	1489575	−71.9	205946	−62.4	1157
10	112053	−41.0	97935	−75.9	1469442	−71.3	206003	−62.5	1163
11	112053	−40.4	94644	−75.4	1450710	−71.3	203486	−62.0	1157
12	112053	−41.3	98867	−76.8	1530771	−72.8	204793	−62.4	1149
13	112053	−40.8	94884	−75.4	1441953	−70.8	206064	−62.3	1161
14	112053	−41.0	98394	−76.6	1510638	−72.6	205732	−62.7	1147
15	112053	−40.7	94264	−75.2	1482856	−71.3	204249	−62.0	1163
16	112053	−41.0	97181	−76.1	1506839	−73.0	205707	−62.6	1150
17	112053	−40.6	95659	−75.8	1467540	−71.9	204496	−62.1	1156
18	112053	−41.4	96728	−76.2	1471799	−72.1	208705	−63.4	1144
19	112053	−40.8	94990	−75.5	1413735	−70.6	207759	−62.4	1161
20	112053	−40.9	95524	−75.8	1459305	−71.6	206216	−62.7	1149
21	112053	−41.4	98716	−76.7	1495316	−72.3	207658	−62.8	1149
22	112053	−41.3	98325	−76.3	1465781	−71.4	206091	−62.3	1160
23	112053	−41.1	97718	−76.4	1477553	−72.4	206640	−62.8	1148
24	112053	−41.2	98202	−76.2	1514816	−71.7	204939	−62.5	1156
25	112053	−40.9	96968	−76.0	1502011	−72.2	205282	−62.3	1157
Average	112053	−41.0	96738	−76.0	1476396	−71.8	205747	−62.4	1154

Tax effects across households:

Average tax change per vehicle for one-vehicle HHs	= $121.27
Average tax change per vehicle for two-vehicle HHs	= $135.31
Average tax change per household for two-vehicle HHs	= $270.62
Average tax change per vehicle for multi-vehicle HHs	= $104.24
Average tax change per household for multi-vehicle HHs	= $360.40

Table 4.2 TIERS model: travel and emissions summary using the Walls et al. elasticities

Emissions tax rates ($/g): HC = $0.0165, CO = $0.00022, NO_x = $0.0088

One-vehicle household:	own-price elasticity	=	−0.13
Two-vehicle household:	own-price elasticity	=	−0.52
	cross-price elasticity	=	0.20
Multi-vehicle household:	own-price elasticity	=	−0.92
	cross-price elasticity	=	−0.04

Iteration	VMT base	% change	HC base	% change	CO base	% change	NO_x base	% change	Revenue ($mil)
1	112053	−22.9	95967	−52.4	1495330	−47.9	204613	−38.2	2037
2	112053	−22.5	95123	−52.2	1447596	−47.0	203202	−38.1	2027
3	112053	−22.7	96289	−53.0	1435247	−47.1	204679	−38.5	2021
4	112053	−22.8	98231	−53.2	1489415	−48.4	204870	−38.7	2033
5	112053	−22.9	96787	−52.9	1445413	−48.0	205984	−38.5	2032
6	112053	−22.9	98064	−52.5	1501377	−47.4	203911	−38.5	2045
7	112053	−23.2	97449	−53.4	1480044	−48.1	209683	−39.3	2038
8	112053	−22.8	96524	−52.7	1464844	−48.0	206972	−39.0	2031
9	112053	−22.8	95015	−53.1	1489575	−47.8	205946	−38.8	2015
10	112053	−22.8	97935	−53.0	1469442	−47.0	206003	−38.3	2050
11	112053	−22.6	94644	−52.8	1450710	−48.4	203486	−38.5	2004
12	112053	−23.1	98867	−52.9	1530771	−47.9	204793	−38.3	2054
13	112053	−22.8	94884	−51.6	1441953	−47.0	206064	−38.6	2040
14	112053	−23.0	98394	−54.3	1510638	−49.1	205732	−38.8	2019
15	112053	−22.7	94264	−52.5	1482856	−48.4	204249	−38.3	2017
16	112053	−23.1	97181	−53.3	1506839	−50.2	205707	−38.6	2024
17	112053	−22.7	95659	−52.6	1467540	−47.9	204496	−38.3	2027
18	112053	−23.2	96728	−52.7	1471799	−48.5	208705	−39.3	2036
19	112053	−22.8	94990	−52.9	1413735	−46.9	207759	−38.8	2024
20	112053	−22.7	95524	−52.5	1459305	−48.2	206216	−38.3	2035
21	112053	−23.3	98716	−54.0	1495316	−49.1	207658	−39.0	2031
22	112053	−23.0	98325	−52.9	1465781	−47.6	206091	−38.3	2053
23	112053	−23.0	97718	−53.2	1477553	−49.1	206640	−39.4	2022
24	112053	−23.0	98202	−52.8	1514816	−48.2	204939	−38.4	2048
25	112053	−22.8	96968	−53.2	1502011	−48.4	205282	−38.4	2033
Average	112053	−22.9	96738	−52.9	1476396	−48.1	205747	−38.6	2032

Tax effects across households:

Average tax change per vehicle for one-vehicle HHs	= $355.89
Average tax change per vehicle for two-vehicle HHs	= $238.25
Average tax change per household for two-vehicle HHs	= $476.51
Average tax change per vehicle for multi-vehicle HHs	= $107.24
Average tax change per household for multi-vehicle HHs	= $370.77

Applying the TIERS percentages to the CARB data, emissions reductions would be 204–293 TPD for ROG, 1227–1841 TPD for CO and 108–172 TPD for NO_x. These figures represent 33–47 per cent of the 1994 ROG reduction target from all sources specified in the 1991 *Air Quality Management Plan* (SCAQMD, 1991), 49–72 per cent of the CO target and 34–55 per cent of the NO_x target.

The ROG reduction target exceeds total ROG emissions from all on-road mobile sources, including heavy duty trucks, buses and vans, by 19 TPD. CO and NO_x targets could be met through on-road mobile source reduction alone. These results argue strongly that this tax could play a substantial role in achieving the South Coast's goal of 'clean air, at the lowest possible cost'.

Even if one exempted pre-1976 (classic) vehicles from emissions testing and taxation, emissions reduction would still be significant under this tax. Pre-1976 vehicles produced only 18 per cent of HC, 14.7 per cent of CO and 15.3 per cent of NO_x emissions in the South Coast. Leaving their emissions unchanged while inducing the newer vehicles to abate would result in area-wide reductions of 57.6 per cent of HC, 57.2 per cent of CO and 46.9 per cent of NO_x emissions based on the author's elasticities.

It is important to note that the results in Tables 4.1 and 4.2 are consistent with each other, although the price elasticities are strikingly different. In both scenarios, emissions reduction exceeds VMT reduction, implying that the tax would encourage higher-emitting vehicles to be driven proportionately fewer miles than lower-emitting vehicles. This is consistent with the tax design in which there is no charge for driving *per se*, only for producing emissions.

Table 4.1 shows a greater reduction in VMT and emissions than the less elastic Walls et al. models which form the basis of Table 4.2. Tax revenues per vehicle and per household are higher when VMT price elasticity is lower because fewer miles will be reduced in that case. (The tax is assumed to be paid on the miles the household drives after the tax is imposed.)

When comparing the results in Tables 4.1 and 4.2, it is interesting to note that, when using the author's own elasticities, the tax paid per vehicle is relatively constant ($121.27, $135.31 and $104.24 for one-, two- and three-or-more-vehicle households, respectively), whereas, when using the Walls et al. elasticities, the tax paid per household is relatively close ($355.89, $476.51 and $370.77, respectively). To the extent the household faces a single budget constraint, one would expect the household-level expenditures on travel to be fairly constant. However, to the extent each vehicle has its own primary driver with an individual (not household) budget constraint, it might be expected that the tax per vehicle would be more constant. Either decision-making unit may be correct, depending on the household's own circumstances. The recent work of Golub et al. (1994) suggests that the individual driver may be the more important decision-maker in determining individual vehicle VMT.

TAX EQUITY ISSUES

The effect of the emissions tax on households of different income levels has been examined to estimate the equity effects of the tax. Households have been divided into income quintiles and TIERS has been run across the one-vehicle, two-vehicle and three-or-more-vehicle households in each quintile. Quintiles were defined as follows:

Quintile 1: income up to $20000;
Quintile 2: income between $20001 and $35000;
Quintile 3: income between $35001 and $50000;
Quintile 4: income between $50001 and $75000;
Quintile 5: income above $75000.

The results, detailed in Table 4.3, indicate that the tax paid per vehicle is greater for the two highest quintiles and nearly equal across the three lower quintiles. VMT declines inversely to income, with the lowest income quintile experiencing a 42 per cent reduction in travel compared with a 35 per cent reduction for the highest income quintile.

The results in Table 4.3 are consistent with the pattern of vehicle holdings across income groups. As Table 4.4 shows, lower-income households own fewer, but older, vehicles on average than higher-income households. Pre-1981 vehicles account for 38 per cent of the lowest income quintile's holdings but only 17 per cent of the highest income quintile's holdings. On average, these older vehicles emit more per mile than newer vehicles but are driven fewer miles. This tends to flatten the dollar effects of the tax across households. The lowest income quintile owns only 13.8 per cent of the area's vehicles but 18.4 per cent of the pre-1976 and 19.8 per cent of the pre-1981 vehicles. By contrast, the highest income quintile owns 19 per cent of the area's vehicles but only 10.5 per cent of pre-1976 and 11.8 per cent of pre-1981 vehicles. The tax would clearly be regressive.

The Suits index (Suits, 1977) can be used to estimate the percentage of a tax burden that falls on each income group. Mathematically, the index, S, equals $(K-L)/K$, where K is the area of the triangle OAB in Figure 4.2 and L is the area OABC (Suits, 1977). When OABC is above OB, as in Figure 4.2, the tax is regressive.

The x-axis in Figure 4.2 represents the cumulative percentage of income, y, and the y-axis represents the corresponding cumulative percentage of the tax paid, $T_x(y)$. The area under the curve corresponding to a given tax, x, is:

$$L_x = \int_0^{100} T_x(y)\,dy$$

Table 4.3 Effects of the emissions tax by income quintile

	Tax ($) per vehicle	Tax ($) per household	% change in VMT
QUINTILE 1			−41.8
One-vehicle households	117.20	117.20	
Two-vehicle households	126.20	252.40	
Three-or-more-vehicle households	90.65	311.24	
QUINTILE 2			−38.4
One-vehicle households	127.06	127.06	
Two-vehicle households	124.73	249.45	
Three-or-more-vehicle households	89.71	310.44	
QUINTILE 3			−36.8
One-vehicle households	118.35	118.35	
Two-vehicle households	121.99	243.99	
Three-or-more-vehicle households	97.87	338.77	
QUINTILE 4			−35.1
One-vehicle households	130.84	130.84	
Two-vehicle households	128.19	256.39	
Three-or-more-vehicle households	105.49	364.23	
QUINTILE 5			−34.7
One-vehicle households	133.94	133.94	
Two-vehicle households	134.90	269.80	
Three-or-more-vehicle households	112.75	390.49	

Table 4.4 Vehicle holdings by income quintile

	Pre-1981 vehicles as % of quintile holdings	1981–85 vehicles as % of quintile holdings	1986–92 vehicles as % of quintile holdings	Average number of vehicles per household
Quintile 1	38.4	31.8	29.9	1.48
Quintile 2	33.7	27.1	39.2	1.79
Quintile 3	27.1	26.4	46.5	2.04
Quintile 4	22.0	25.0	52.9	2.31
Quintile 5	16.6	24.0	59.3	2.44

Suits claimed L_x can be approximated by using discrete measures of the cumulative tax burden and cumulative income:

$$L_x = \sum_i (1/2)[T_x(y_i) + T_x(y_{i-1})](y_i - y_{i-1})$$

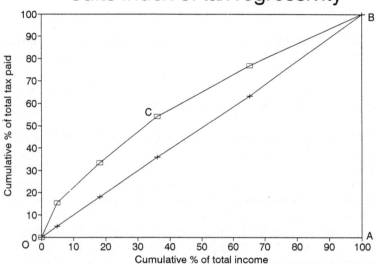

Figure 4.2 The Suits index for the mobile source emissions tax

Table 4.5 shows the calculation of the Suits index for the mobile source emissions tax.

Table 4.5 Data used to calculate the Suits index for the emissions tax

Quintile	Tax paid ($mil)	% of tax revenue	Cumulative % of tax	Mean income	% of total income	Cumulative % of income	Quintile L value
1	178	15.5	15.5	12500	4.9	4.9	38.0
2	207	18.0	33.5	27500	13.3	18.2	325.9
3	238	20.7	54.2	42500	17.8	36.0	780.5
4	261	22.6	76.8	62500	29.3	63.3	1919.2
5	267	23.2	100.0	105000	34.7	100.0	3067.5

L_x is the sum of the 'quintile L value' column and equals 6131.
K, the area of the triangle OAB in Figure 4.2, has a constant value of 5000 since the base and height of the triangle always equal 100. Based on the data in Table 4.5, $S = 1-(6131/5000) = -0.226$. This is highly regressive.
Suits' 1977 study calculated sales and excise taxes to have an index value of

−0.15 in 1970, and personal property and motor vehicle taxes to have an index value of −0.09 (Suits, 1977, p. 750). Even allowing for the temporal variation between 1970 and 1990, the comparative degree of regressivity of the mobile source emissions tax is significant.

REVENUE ISSUES

The TIERS model shows that the emissions tax would yield approximately $1160 million in tax revenue after *ex ante* VMT is reduced by approximately 41 per cent. The question arises as to whether this tax can be made revenue-neutral. If revenue neutrality can be achieved, a further gain can be achieved if the tax to be reduced is less efficient or more distortionary than the emissions tax which would replace it.

Although the California state gasoline tax appears initially to be a candidate for such a reduction, it would be a questionable choice for two reasons:

1. It would not be possible to reduce the gasoline tax in the South Coast region enough to offset the revenue generated by the emissions tax because the average emissions tax exceeds the average fuel tax by nearly one cent per mile. The emissions tax averages 1.73 cents per mile. The California state gasoline tax was 15 cents per gallon in 1991, which equates to 0.74 cents per mile based on the CEC's fleet average fuel economy for the South Coast region of 20.26 MPG (0.15/20.26=0.0074). It would be possible to reduce gasoline taxes on a state-wide basis to achieve revenue neutrality, assuming the emissions tax is levied only in the four counties of the South Coast Air Basin. Gasoline and diesel fuel tax revenues yielded $2002 million in 1991 (AAMA, 1993, p. 80). The neutrality of an offsetting gasoline tax reduction would, therefore, depend upon the scope of the taxing jurisdiction.
2. Although it is possible to achieve revenue neutrality state-wide, a reduction in gasoline taxes would conflict directly with the goal of the emissions tax, namely to reduce emissions in the South Coast region. Any decrease in the gasoline tax would encourage increased VMT and increased emissions state-wide. This would hurt the South Coast as well as other regions, such as San Diego and the San Francisco Bay area, which must reduce emissions to conform with the NAAQS.

The TIERS model has been run to examine the partial effect of a gasoline tax reduction of 15 cents per gallon and the joint effects of the emissions tax combined with a gasoline tax reduction. Tables 4.6 and 4.7 show these impacts on VMT and emissions. Table 4.8 summarizes the results.

Table 4.6 TIERS model: travel and emissions effects of a gasoline tax reduction of $0.15 per gallon

Emissions tax rates ($/g): HC = $0.00, CO = $0.00, NO_x = $0.00

One-vehicle household:	own-price elasticity	=	−0.85189
Two-vehicle household:	own-price elasticity	=	−0.91559
	cross-price elasticity	=	0.13783
Multi-vehicle household:	own-price elasticity	=	−0.94096
	cross-price elasticity	=	0.30115

Iteration	VMT base	% change	HC base	% change	CO base	% change	NO_x base	% change	Revenue ($mil)
1	112053	10.5	95967	10.6	1495330	10.6	204613	10.6	−775
2	112053	10.5	95123	10.6	1447596	10.6	203202	10.6	−775
3	112053	10.5	96289	10.6	1435247	10.6	204679	10.6	−775
4	112053	10.5	98231	10.6	1489415	10.6	204870	10.6	−775
5	112053	10.5	96787	10.6	1445413	10.6	205984	10.6	−775
6	112053	10.5	98064	10.6	1501377	10.6	203911	10.6	−775
7	112053	10.5	97449	10.6	1480044	10.6	209683	10.6	−775
8	112053	10.5	96524	10.6	1464844	10.6	206972	10.6	−775
9	112053	10.5	95015	10.6	1489575	10.6	205946	10.6	−775
10	112053	10.5	97935	10.6	1469442	10.6	206003	10.6	−775
11	112053	10.5	94644	10.6	1450710	10.6	203486	10.6	−775
12	112053	10.5	98867	10.6	1530771	10.6	204793	10.6	−775
13	112053	10.5	94884	10.6	1441953	10.6	206064	10.6	−775
14	112053	10.5	98394	10.6	1510638	10.6	205732	10.6	−775
15	112053	10.5	94264	10.6	1482856	10.6	204249	10.6	−775
16	112053	10.5	97181	10.6	1506839	10.6	205707	10.6	−775
17	112053	10.5	95659	10.6	1467540	10.6	204496	10.6	−775
18	112053	10.5	96728	10.6	1471799	10.6	208705	10.6	−775
19	112053	10.5	94990	10.6	1413735	10.6	207759	10.6	−775
20	112053	10.5	95524	10.6	1459305	10.6	206216	10.6	−775
21	112053	10.5	98716	10.6	1495316	10.6	207658	10.6	−775
22	112053	10.5	98325	10.6	1465781	10.6	206091	10.6	−775
23	112053	10.5	97718	10.6	1477553	10.6	206640	10.6	−775
24	112053	10.5	98202	10.6	1514816	10.6	204939	10.6	−775
25	112053	10.5	96968	10.6	1502011	10.6	205282	10.6	−775
Average	112053	10.5	96738	10.6	1476396	10.6	205747	10.6	−775

Tax effects across households:

Average tax change per vehicle for one-vehicle HHs	=	−$ 94.11
Average tax change per vehicle for two-vehicle HHs	=	−$ 92.72
Average tax change per household for two-vehicle HHs	=	−$185.44
Average tax change per vehicle for multi-vehicle HHs	=	−$ 91.44
Average tax change per household for multi-vehicle HHs	=	−$316.16

Table 4.7 TIERS model: effects of a gasoline tax reduction of $0.15 per gallon combined with the emissions tax

Emissions tax rates ($/g): HC = $0.0165, CO = $0.00022, NO$_x$ = $0.0088

One-vehicle household:	own-price elasticity	=	−0.85189
Two-vehicle household:	own-price elasticity	=	−0.91559
	cross-price elasticity	=	0.13783
Multi-vehicle household:	own-price elasticity	=	−0.94096
	cross-price elasticity	=	0.30115

Iteration	VMT base	% change	HC base	% change	CO base	% change	NO$_x$ base	% change	Revenue ($mil)
1	112053	−31.6	95967	−69.6	1495330	−65.3	204613	−54.2	646
2	112053	−31.1	95123	−69.6	1447596	−64.4	203202	−54.1	637
3	112053	−31.4	96289	−69.8	1435247	−64.3	204679	−54.4	640
4	112053	−31.5	98231	−71.0	1489415	−66.1	204870	−54.7	623
5	112053	−31.5	96787	−70.3	1445413	−65.3	205984	−53.9	647
6	112053	−31.8	98064	−70.7	1501377	−66.0	203911	−54.4	630
7	112053	−32.0	97449	−70.7	1480044	−65.6	209683	−55.1	638
8	112053	−31.5	96524	−70.3	1464844	−65.5	206972	−54.7	636
9	112053	−31.4	95015	−69.6	1489575	−65.4	205946	−54.5	640
10	112053	−31.6	97935	−70.0	1469442	−64.7	206003	−54.5	648
11	112053	−31.0	94644	−69.3	1450710	−64.7	203486	−54.0	641
12	112053	−31.9	98867	−71.1	1530771	−66.5	204793	−54.5	629
13	112053	−31.4	94884	−69.4	1441953	−64.1	206064	−54.3	647
14	112053	−31.7	98394	−71.0	1510638	−66.3	205732	−54.9	626
15	112053	−31.4	94264	−69.3	1482856	−64.8	204249	−54.0	644
16	112053	−31.7	97181	−70.3	1506839	−66.7	205707	−54.6	634
17	112053	−31.2	95659	−69.9	1467540	−65.4	204496	−54.0	639
18	112053	−32.0	96728	−70.3	1471799	−65.5	208705	−55.5	629
19	112053	−31.4	94990	−69.7	1413735	−64.0	207759	−54.5	645
20	112053	−31.5	95524	−69.9	1459305	−65.1	206216	−54.7	633
21	112053	−32.1	98716	−71.0	1495316	−65.9	207658	−54.9	634
22	112053	−31.9	98325	−70.4	1465781	−64.7	206091	−54.3	648
23	112053	−31.8	97718	−70.7	1477553	−66.2	206640	−54.9	628
24	112053	−31.8	98202	−70.4	1514816	−65.2	204939	−54.5	640
25	112053	−31.5	96968	−70.2	1502011	−65.9	205282	−54.3	640
Average	112053	−31.6	96738	−70.2	1476396	−65.3	205747	−54.5	638

Tax effects across households:

Average tax change per vehicle for one-vehicle HHs	=	$ 97.88
Average tax change per vehicle for two-vehicle HHs	=	$102.86
Average tax change per household for two-vehicle HHs	=	$205.72
Average tax change per vehicle for multi-vehicle HHs	=	$ 85.41
Average tax change per household for multi-vehicle HHs	=	$295.31

Table 4.8 Comparison of the effects of emissions and gasoline taxes

	VMT % change	HC % change	CO % change	NO_x % change	Revenue change ($million)
Emissions tax only	−41.0	−76.0	−71.8	−62.4	+1154
Gasoline tax reduced 15 cents per gallon	+10.5	+10.6	+10.6	+10.6	− 775
Emissions and gasoline tax combined	−31.6	−70.2	−65.3	−54.5	+ 638

As Table 4.8 shows, combining the emissions tax with a reduction in the gasoline tax would decrease the emissions savings by 7.6 per cent for HC, 9.1 per cent for CO and 12.7 per cent for NO_x compared with the emissions tax alone. While this is a non-trivial set-back for air quality improvement, the fact that the emissions tax would replace a tax which is less efficient for emissions reduction produces a net improvement in the overall efficiency of the tax system.

Revenue neutrality could also be achieved without undermining emissions reduction by reducing motor vehicle registration and licensing fees. State-wide registration and licensing fees totalled $2078.1 million in 1991, at the rate of $24 per vehicle for registration and $98 per vehicle for licensing (Federal Highway Administration, 1992b, p. 43). The South Coast region had 5.7 million registered passenger cars (excluding light trucks, sport-utility vehicles, vans and minivans), approximately one-third of the state-wide registration of 17.2 million passenger vehicles. If passenger vehicle licensing were reduced to $31, the $67 savings per vehicle multiplied by the 17.2 million vehicles would reduce revenue by $1152.4 million, approximately equalling the emissions tax revenue.

The emissions tax revenue could also be used to reduce the portion of the general sales tax that is dedicated to transportation. The transportation-dedicated sales tax rate ranges from 1–1.5 per cent. The 1990–91 California state budget estimated the revenue from the dedicated sales tax to be $2.2 billion (Cameron, 1991, p. 56).

Reducing the sales tax would be preferable to reducing either the gasoline tax or the vehicle licensing and registration fees for several reasons:

- general sales taxes are regressive;
- the transportation benefits received by the person who pays the tax are independent of the amount of tax paid, making the sales tax an inefficient instrument for pricing transportation services;
- VMT and emissions would not increase as they would under a reduction

in the gasoline tax;

• the size of the vehicle fleet would not be affected as it would be under a reduction in registration and licensing fees.

Although reducing the fixed cost of vehicle ownership has no effect on vehicle operating costs, empirical studies agree that VMT growth is strongly correlated with the size of the vehicle fleet.[3] A reduction in licensing or registration fees would encourage increased vehicle ownership and, consequently, increased VMT. This would lead to increased emissions in the region.

Any revenue-neutral plan would require careful monitoring over time to balance the emissions tax revenue with the chosen offsets. As older vehicles are retired from the fleet, average emissions per mile will decline. As the population and vehicle fleet size grow, VMT will increase. The VMT growth rate is expected to be less than the rate of emissions reduction, leading to a net reduction in mobile source emissions in the South Coast (SCAQMD, 1991) and lower tax revenue over time.

Although emissions variability is likely to persist, it is likely that rates will tend to approach convergence at a lower level than current fleet averages. If emissions rates reached a long-run equilibrium, the emissions tax would become a *de facto* VMT tax, as all vehicles would produce the same emissions per mile. Under this scenario, tax revenue would increase with VMT growth. Neither situation would produce a steady-state emissions tax revenue over time.

SUMMARY

Simulations from the TIERS model show that a mobile source emissions tax could have a significant impact on VMT and emissions in the South Coast region. The tax would also generate substantial revenues, some of which could be offset by reductions in other taxes to produce revenue neutrality.

Although lower-income households would be expected to pay fewer dollars than upper-income households, either on a per-vehicle or per-household basis, the emissions tax would be regressive. This is consistent with the general regressivity of consumption taxes. The high degree of regressivity, as measured by the Suits index, indicates that the use of this tax might require a careful analysis of the most appropriate use of the revenue to ensure social equity.

APPENDIX 4.1

This appendix contains detailed results from Iteration 1 of Table 4.1, the TIERS model run with the author's elasticities and tax rates. The following terms are used:

> *BASELINE*: the weighted VMT before the tax is imposed;
> *ADJUSTED*: simulated value reflecting the effect of the tax;
> *HC (CO, NO$_x$) BASE*: emissions before the tax is imposed;
> *HC (CO, NO$_x$) NEW*: emissions after the tax is imposed;
> *TAX $MIL*: tax revenue resulting from *ADJUSTED VMT* and emissions.

Table 4.A.1 One-vehicle households' VMT (millions)

Tax rates: HC = $0.0165, CO = $0.00022, NO$_x$ = $0.0088
Own-price elasticity = −0.85189

Model year	Number of vehicles	Baseline VMT	Adjusted VMT
pre-1972	32	351.0	112.6
1972	7	91.5	17.8
1973	8	108.2	24.4
1974	8	94.8	24.7
1975	3	42.7	14.0
1976	17	241.8	73.3
1977	14	207.5	49.9
1978	18	290.5	74.8
1979	27	445.6	128.8
1980	29	473.5	141.2
1981	34	584.2	212.1
1982	31	615.1	150.0
1983	34	720.7	267.3
1984	55	1239.4	496.3
1985	72	1612.1	847.6
1986	52	1288.4	666.5
1987	56	1351.9	965.9
1988	59	1545.6	1092.8
1989	78	2178.3	1622.6
1990	46	1281.5	1070.3
1991	22	649.5	531.9
Total	702	15413.8	8585.1

Table 4.A.2 One-vehicle households' emissions: baseline and adjusted (metric tonnes)

Tax rates: HC = $0.0165, CO = $0.00022, NO_x = $0.0088
Own-price elasticity = −0.85189

Model year	HC base	HC new	CO base	CO new	NO_x base	NO_x new	Tax ($mil)
pre-1972	892.9	150.6	8925.2	2262.9	1344.7	336.6	5.9
1972	226.8	36.6	2741.9	534.6	453.0	84.5	1.5
1973	343.7	30.3	5866.1	558.4	570.8	151.6	2.0
1974	265.3	32.3	1863.1	357.6	436.6	111.6	1.6
1975	151.7	17.1	1620.7	104.9	129.0	53.5	0.8
1976	770.9	108.9	6912.8	1842.9	1003.1	231.3	4.2
1977	382.0	55.8	5007.8	890.9	1075.5	143.1	2.4
1978	648.8	81.2	7495.4	1561.8	1202.0	228.1	3.7
1979	1042.1	106.4	13727.0	1679.7	1670.1	325.2	5.0
1980	863.5	85.0	12357.1	1690.7	1363.5	283.8	4.3
1981	912.5	133.5	17919.7	2756.0	1503.1	393.2	6.3
1982	991.7	105.5	19621.6	1712.2	1660.2	260.2	4.4
1983	1009.0	149.1	15475.3	1833.2	1628.9	399.1	6.4
1984	1189.2	263.3	19542.5	3886.4	2920.3	771.0	12.0
1985	1055.3	360.7	22361.7	7735.4	2547.0	1141.3	17.7
1986	951.2	227.3	14591.7	4170.1	2306.5	912.3	12.7
1987	386.1	241.0	5145.8	3404.2	1609.9	1050.5	14.0
1988	537.2	312.6	8006.3	5177.9	1759.0	1068.9	15.7
1989	616.2	352.9	13278.5	7122.3	2135.9	1485.1	20.5
1990	177.5	140.0	3820.7	3054.8	970.3	771.9	9.8
1991	126.0	89.5	2683.0	1987.5	454.9	341.6	4.9
Total	13539.4	3079.6	208963.8	54324.2	28744.2	10544.6	155.6

Table 4.A.3 Two-vehicle households' VMT (millions)

Tax rates: HC = $0.0165, CO = $0.00022, NO_x = $0.0088
Own-price elasticity   = −0.91559
Cross-price elasticity =   0.13783

Model year	Number of vehicles	Baseline VMT	Adjusted VMT
pre-1972	280	2202.9	525.7
1972	46	376.1	145.0
1973	50	472.1	129.6
1974	42	391.1	116.9
1975	42	385.1	128.0
1976	78	727.4	228.5
1977	110	1127.2	378.4
1978	116	1188.4	313.1
1979	166	1756.0	538.8
1980	136	1609.3	522.6
1981	110	1438.0	444.4
1982	194	2568.0	870.9
1983	186	2710.9	1117.8
1984	312	4659.7	2203.1
1985	378	6034.8	3314.3
1986	396	6584.1	4060.0
1987	374	6855.2	4796.1
1988	332	6215.6	4822.0
1989	408	7901.3	6622.6
1990	378	7603.8	6789.0
1991	238	4843.5	4300.2
Total	4372	67650.2	42467.5

Table 4.A.4 Two-vehicle households' emissions: baseline and adjusted (metric tonnes)

Tax rates: HC = \$0.0165, CO = \$0.00022, NO_x = \$0.0088
Own-price elasticity   = −0.91559
Cross-price elasticity =   0.13783

Model year	HC base	HC new	CO base	CO new	NO_x base	NO_x new	Tax (\$mil)
pre-1972	4972.4	907.9	67224.9	14934.6	9498.3	2059.1	36.3
1972	732.0	164.0	9374.8	2668.1	1443.5	439.1	7.2
1973	1322.6	136.1	15151.3	2168.7	2071.3	426.3	6.5
1974	894.4	154.8	13079.0	3069.5	1653.0	378.4	6.6
1975	769.6	152.4	8574.6	2454.7	1680.9	330.9	6.0
1976	1824.1	297.7	23431.4	5148.8	2898.8	698.0	12.2
1977	2834.4	476.7	31188.0	6395.2	4501.6	1098.6	18.9
1978	3258.6	378.4	43459.8	5075.7	4615.1	1021.2	16.3
1979	3634.4	619.9	44316.5	8944.1	7610.8	1818.1	28.2
1980	2849.4	401.8	39192.2	7055.9	5294.8	1169.3	18.5
1981	2559.6	268.2	61881.0	5186.8	3716.7	837.4	12.9
1982	3755.2	538.4	69377.7	10452.1	6425.5	1493.0	24.3
1983	3107.4	594.8	51843.1	9925.6	6450.1	2022.4	29.8
1984	4354.2	1168.3	76604.0	21176.4	10090.0	3509.5	54.8
1985	5080.7	1573.3	84269.0	27997.6	10668.4	4385.6	70.7
1986	4763.0	1478.0	76451.5	23881.3	9767.0	4828.6	72.1
1987	3476.2	1424.3	64566.0	26314.3	8865.4	5204.5	75.1
1988	1954.3	1287.5	33999.3	22874.6	6828.6	4703.2	67.7
1989	1894.4	1327.3	35774.1	25265.8	7156.3	5556.6	76.4
1990	1520.0	1012.8	32699.8	23603.1	5727.6	4788.9	64.0
1991	952.4	781.9	20432.8	16873.7	3233.9	2737.6	40.7
Total	56509.2	15144.2	902890.9	271466.5	120197.8	49506.7	745.3

Table 4.A.5 Three-or-more-vehicle households' VMT (millions)

Tax rates: HC = \$0.0165, CO = \$0.00022, NO_x = \$0.0088
Own-price elasticity   = -0.94096
Cross-price elasticity =   0.30115

Model year	Number of vehicles	Baseline VMT	Adjusted VMT
pre-1972	226	1920.2	410.1
1972	22	210.9	30.2
1973	22	209.3	37.7
1974	33	323.3	73.2
1975	21	227.2	32.0
1976	36	384.1	76.6
1977	50	568.5	135.6
1978	52	594.6	117.2
1979	72	887.0	219.8
1980	62	796.9	161.4
1981	55	766.8	251.1
1982	60	875.7	247.4
1983	65	1033.9	336.0
1984	97	1573.9	581.5
1985	128	2346.9	1093.1
1986	135	2524.1	1364.5
1987	145	2806.6	1708.6
1988	139	2840.7	1946.5
1989	150	3138.5	2277.0
1990	133	2998.2	2408.0
1991	88	1961.4	1546.1
Total	1791	28988.8	15078.6

62   *Taxing Automobile Emissions for Pollution Control*

Table 4.A.6 Three-or-more-vehicle households' emissions: baseline and adjusted (metric tonnes)

Tax rates: HC = $0.0165, CO = $0.00022, NO_x = $0.0088
Own-price elasticity   = −0.94096
Cross-price elasticity =   0.30115

Model year	HC base	HC new	CO base	CO new	NO_x base	NO_x new	Tax ($mil)
pre-1972	4366.1	532.8	58590.7	8960.3	7710.4	1096.1	21.0
1972	484.8	32.1	5518.0	567.8	1102.3	102.2	1.6
1973	438.2	39.0	5537.8	618.5	1034.1	130.9	1.9
1974	750.8	93.2	8992.2	1261.0	1216.1	181.0	3.4
1975	645.8	41.9	6550.5	680.3	945.8	86.0	1.6
1976	939.7	78.9	12640.2	1653.0	1652.1	245.4	3.8
1977	1467.3	159.6	17925.3	2549.3	2246.5	381.0	6.5
1978	1581.6	133.7	18407.3	2055.0	2355.9	360.2	5.8
1979	1743.8	233.5	24192.8	2746.6	3332.2	589.9	9.6
1980	1654.2	104.1	21587.8	1789.7	2759.5	370.3	5.4
1981	809.9	169.8	13963.2	3182.0	2161.0	457.9	7.5
1982	1113.4	121.2	13921.2	2097.0	2468.4	460.0	6.5
1983	1221.8	151.0	22267.7	2374.5	2455.5	637.5	8.6
1984	1303.2	285.6	22502.9	4929.1	3482.0	893.2	13.7
1985	2016.4	453.5	33236.8	8558.2	3872.2	1385.3	21.6
1986	1540.0	491.1	23464.7	7626.7	3641.9	1530.1	23.2
1987	1302.2	473.9	20440.4	8423.7	3602.7	1801.3	25.5
1988	926.4	549.3	15691.4	9567.2	2989.5	1825.2	27.2
1989	730.2	447.9	16535.9	10320.5	2983.2	1967.0	27.0
1990	509.3	362.2	12525.9	9113.2	2210.2	1638.3	22.4
1991	373.7	258.1	8982.6	5968.0	1449.7	1103.6	15.3
Total	25918.6	5212.4	383475.5	95041.8	55671.2	17314.3	259.3

Table 4.A.7 VMT change induced by the emissions tax (millions of miles)

Model year	Base VMT	Adjusted VMT	VMT decline	% change VMT
pre-1972	4474.3	1173.7	3300.6	−73.8
1972	678.5	192.9	485.5	−71.6
1973	789.5	191.7	597.8	−75.7
1974	809.1	214.8	594.3	−73.5
1975	654.9	174.1	480.8	−73.4
1976	1353.3	378.4	974.9	−72.0
1977	1903.2	563.9	1339.3	−70.4
1978	2073.5	505.1	1568.5	−75.6
1979	3088.6	887.4	2201.2	−71.3
1980	2879.7	825.1	2054.5	−71.3
1981	2789.1	907.6	1881.4	−67.5
1982	4058.7	1268.3	2790.4	−68.8
1983	4465.5	1721.2	2744.3	−61.5
1984	7473.0	3281.0	4192.0	−56.1
1985	9993.8	5255.0	4738.8	−47.4
1986	10396.7	6091.0	4305.6	−41.4
1987	11013.6	7470.7	3543.0	−32.2
1988	10601.9	7861.3	2740.5	−25.8
1989	13218.0	10522.2	2695.8	−20.4
1990	11883.6	10267.3	1616.3	−13.6
1991	7454.4	6378.3	1076.1	−14.4
Total	112052.8	66131.2	45921.7	−41.0

Table 4.A.8 HC emissions change induced by the emissions tax (metric tonnes)

Model year	Base HC	Adjusted HC	HC decline	% change HC
pre-1972	10231.4	1591.3	8640.1	−84.4
1972	1443.5	232.7	1210.8	−83.9
1973	2104.4	205.3	1899.1	−90.2
1974	1910.5	280.2	1630.2	−85.3
1975	1567.1	211.4	1355.6	−86.5
1976	3534.7	485.4	3049.3	−86.3
1977	4683.8	692.1	3991.7	−85.2
1978	5489.0	593.3	4895.7	−89.2
1979	6420.3	959.8	5460.4	−85.1
1980	5367.1	590.9	4776.1	−89.0
1981	4282.0	571.4	3710.6	−86.7
1982	5860.3	765.0	5095.3	−86.9
1983	5338.2	894.9	4443.3	−83.2
1984	6846.6	1717.2	5129.4	−74.9
1985	8152.4	2387.5	5764.9	−70.7
1986	7254.2	2196.4	5057.7	−69.7
1987	5164.4	2139.2	3025.2	−58.6
1988	3417.9	2149.4	1268.5	−37.1
1989	3240.8	2128.1	1112.7	−34.3
1990	2206.7	1515.0	691.7	−31.3
1991	1452.0	1129.5	322.5	−22.2
Total	95967.2	23436.2	72531.0	−75.6

Table 4.A.9 CO emissions change induced by the emissions tax (metric tonnes)

Model year	Base CO	Adjusted CO	CO decline	% change CO
pre-1972	134741.0	26158.1	108582.9	−80.6
1972	17634.7	3770.4	13864.3	−78.6
1973	26555.2	3345.6	23209.6	−87.4
1974	23934.3	4688.1	19246.2	−80.4
1975	16745.8	3240.0	13505.9	−80.7
1976	42984.4	8644.6	34339.7	−79.9
1977	54121.1	9835.5	44285.7	−81.8
1978	69362.5	8692.5	60670.0	−87.5
1979	82236.3	13370.4	68865.9	−83.7
1980	73137.0	10536.3	62600.8	−85.6
1981	93763.9	11124.8	82639.1	−88.1
1982	102920.4	14261.2	88659.2	−86.1
1983	89586.1	14133.3	75452.9	−84.2
1984	118649.5	29991.8	88657.6	−74.7
1985	139867.5	44291.3	95576.3	−68.3
1986	114507.9	35678.1	78829.8	−68.8
1987	90152.2	38142.2	52010.0	−57.7
1988	57697.0	37619.6	20077.5	−34.8
1989	65588.4	42708.5	22879.9	−34.9
1990	49046.4	35771.1	13275.2	−27.1
1991	32098.3	24829.2	7269.2	−22.6
Total	1495330.0	420832.4	1074498.0	−71.9

Table 4.A.10 NO$_x$ emissions change induced by the emissions tax (metric tonnes)

Model year	Base NO$_x$	Adjusted NO$_x$	NO$_x$ decline	% change NO$_x$
pre-1972	18553.3	3564.8	14988.5	−80.8
1972	2998.8	625.8	2373.0	−79.1
1973	3676.2	708.8	2967.4	−80.7
1974	3305.7	671.0	2634.7	−79.7
1975	2755.7	470.4	2285.3	−82.9
1976	5554.1	1174.8	4379.3	−78.8
1977	7823.6	1622.7	6200.9	−79.3
1978	8173.0	1609.5	6563.5	−80.3
1979	12613.1	2733.2	9879.8	−78.3
1980	9417.9	1823.4	7594.5	−80.6
1981	7380.9	1688.5	5692.4	−77.1
1982	10554.1	2213.2	8340.9	−79.0
1983	10534.6	3058.9	7475.6	−71.0
1984	16492.3	5173.6	11318.7	−68.6
1985	17087.7	6912.3	10175.4	−59.5
1986	15715.3	7271.0	8444.3	−53.7
1987	14078.0	8056.3	6021.7	−42.8
1988	11577.1	7597.4	3979.7	−34.4
1989	12275.4	9008.7	3266.7	−26.6
1990	8908.1	7199.1	1709.0	−19.2
1991	5138.4	4182.8	955.6	−18.6
Total	204613.3	77365.5	127247.7	−62.2

NOTES

1. Emissions data from the USEPA I/M 240 testing facility in Hammond, Indiana exist for 1976–91 model year vehicles. 1975 and earlier CalTrans vehicles are assigned an emissions reading from the 1976 distribution. This likely understates the emissions from pre-catalytic converter vehicles but, given the wide distribution of 1976 vehicle emissions and the fact that these older vehicles represent only 12 per cent of the total fleet, it does not appear to be a serious problem.
2. CARB'S EMFAC data include hot and cold start emissions in addition to running exhaust. Reductions attributed to the emissions tax do not include trip start or end emissions.
3. See, for example, Goodwin (1992) and Train (1986).

5. Modelling the Scrappage Effect of the Tax

THEORETICAL MODEL OF VEHICLE SCRAPPAGE

The emissions tax increases a vehicle's operating cost. This in turn affects the household's decisions regarding vehicle ownership, usage and maintenance. Each household is assumed to choose and use its vehicles so as to maximize its utility. The vehicle ownership decision includes both the number and types of vehicles to own. Implicit in the ownership decision is the choice of whether to keep or scrap any vehicle currently owned.

Beginning with Parks (1977), the theory of vehicle scrappage has developed around a simple notion: a vehicle will be scrapped whenever it faces a repair cost that exceeds the vehicle's value. These costs typically result from random events (accidents, natural disasters, malfunction) which leave the vehicle inoperable without the costly repair.

In his model, Parks defined the probability that a car of age a with durability characteristics[1] δ would be scrapped as $\lambda(a,\delta)$. He further defined:

$R(a,\delta)$ = a non-negative random variable representing the size of the repair required of a-year-old cars with durability parameter δ;

$f(R;a,\delta)$ = density function of R;

$q(t)$ = price of the repair services at time t;

$q(t)*(R;a,\delta)$ = the realized repair bill for an a-year-old car in year t;

$P(a,t)$ = the value of a working a-year-old car in year t;

$S(a,t)$ = scrap value of an a-year-old car in year t.

If a vehicle needs repair, the owner makes the repair if and only if $q(t)*(R;a,\delta) < P(a,t) - S(a,t)$, that is, if the net value of the vehicle exceeds the cost of repair. If the inequality is not satisfied, the vehicle would be scrapped. The probability of a vehicle being scrapped is:

$$\lambda(a,\delta) = \int_{[P(a,t)-S(a,t)]/q(t)}^{\infty} f(R;a,\delta)\, dR$$

Scrappage probability increases with vehicle age and relative salvage value, and decreases with relative vehicle value:

$$\frac{\partial \lambda(a,\delta)}{\partial a} > 0$$

$$\frac{\partial \lambda(a,\delta)}{\partial [S(a,t)/q(t)]} > 0$$

$$\frac{\partial \lambda(a,\delta)}{\partial [P(a,t)/q(t)]} < 0$$

Parks claimed it is possible to use the expression for $\lambda(a,\delta)$ directly as the basis for estimating a scrapping rate function by specifying a functional form for the underlying repair distribution and its dependence on age and durability. According to Parks, estimation of the model would be:

> relatively simple because we are interested in a single dichotomous variable, scrapping vs. not scrapping. The data occur in a form in which relative frequencies can be used to approximate the relative scrapping probabilities and the sample sizes used to compute the relative frequencies are very large (Parks, 1977, p. 1103).

Parks' framework can be extended to estimate the probability of vehicle scrappage induced by the emissions tax. A vehicle's emissions control technology (analogous to Parks' durability measure, δ), random failures of the on-board emissions control system, maintenance and vehicle usage jointly determine the vehicle's tax liability. The tax presents the vehicle owner with a similar choice to Parks' realized repair cost: pay the tax to keep the vehicle (legally) operating or scrap the vehicle and save the cost of the tax.[2]

It may appear that the vehicle owner faces three, not two, discrete choices: pay the tax, reduce emissions to reduce the tax or scrap the vehicle. However, the dichotomous nature of the choice is preserved if the owner's decision is viewed (more correctly) as a nested choice: if the minimum of the tax owed or the cost of maintenance plus the new, reduced tax owed is less than or equal to the value of the vehicle, keep the vehicle. If the tax or reduced tax plus maintenance cost exceeds the vehicle value, scrap the vehicle. Following from Parks, we can define:

$E(a,\delta)$ = emissions[3] per mile of a vehicle of age a with emissions control characteristics δ;

$f(E;a,\delta)$ = density function of E;

$E'(a,\delta)$ = emissions per mile of a vehicle of age a with emissions control characteristics δ which has undergone emissions-reducing maintenance;

$f(E';a,\delta)$ = density function of E';
$VMT(a,t)$ = annual miles travelled by a vehicle of age a at time t;
$f(VMT;a,t)$ = density function of VMT;
$q(t)$ = price of emissions-reducing maintenance;
$P(a,t)$ = the value of a working a-year-old car in year t;
$S(a,t)$ = scrap value of an a-year-old car in year t;
c = tax rate per unit of emissions;
$E(a,\delta)*VMT(a,t)*c$ = emissions tax assessed, given a tax rate of c cents per unit of emissions;
$E'(a,\delta)*VMT(a,t)*c$ = post-maintenance emissions tax assessed.

It is expected that $E'(a,\delta)<E(a,\delta)$ but this inequality may be violated.[4] A vehicle will be scrapped if:

$$min[E(a,\delta)*VMT(a,t)*c,(E'(a,\delta)*VMT(a,t)*c+q(t))] > P(a,t)-S(a,t)$$

In this specification, the dichotomous choice is preserved: scrap or keep the vehicle depending upon whether the tax, or the adjusted tax plus maintenance cost, exceeds the net vehicle value.

ESTIMATING THE TAX-INDUCED VEHICLE SCRAPPAGE

Assuming all cost-effective emissions-reducing maintenance has already been performed and further tax-reducing maintenance is not possible, Parks' relative frequency approach can be used to estimate the upper limit of tax-induced vehicle scrappage.

Since the tax rate for each pollutant is fixed, the vehicle faces a tax per mile based on its emissions of HC, CO and NO_x. There are many combinations of emissions and VMT that result in the same tax assessment. For example, a low-emitting vehicle, assessed $0.005 per mile travelling 10000 miles owes the same tax as a high-emitting vehicle assessed $0.20 per mile travelling 250 miles.

The scrappage analysis assesses the probability that the emissions tax will exceed the vehicle value. Taking the vehicle value as fixed, this analysis considers the combinations of emissions tax rates and VMT which would lead to the vehicle being scrapped.

Table 5.1 shows the minimum tax per mile that produces an annual emissions tax in excess of the vehicle value for given combinations of VMT and vehicle value. Any tax per mile above the trigger point value would lead to the vehicle being scrapped. It is important to remember that these trigger points are conditional upon vehicle value and VMT being fixed.

Value can be estimated as the average price of an average-condition used car of a given make, model and year, as quoted in the standard used car guides.[5]

Table 5.1 Tax rate trigger points: emissions-based tax (cents/mile) above which the annual tax exceeds the vehicle value

Vehicle value ($)	VMT: 5000	VMT: 10000	VMT: 15000	VMT: 20000	VMT: 25000	VMT: 30000
500	10.0	5.0	3.3	2.5	2.0	1.7
1000	20.0	10.0	6.7	5.0	4.0	3.3
1500	30.0	15.0	10.0	7.5	6.0	5.0
2000	40.0	20.0	13.3	10.0	8.0	6.7

Table 5.2 shows the distribution of values for pre-1982 vehicles in the South Coast fleet in 1992, as estimated by the CEC.

Table 5.2 Percentage of model year vehicles in a given valuation range: South Coast Air Basin, 1992

Vintage	Up to $500	$501–1000	$1001–1500	$1501–2000
Pre-1977	8.4%	53.5%	1.6%	11.1%
1977		25.7%	36.0%	11.5%
1978		8.3%	53.2%	1.1%
1979		9.8%	43.2%	4.6%
1980			33.5%	23.1%
1981				9.6%

Note: Rows do not sum to 100 per cent because it is expected that vehicles valued above $2000 would be repaired rather than scrapped.

This approach implicitly assumes that future years' taxes are fully discounted, so the current year's tax is the only relevant cost in the scrap-or-keep decision. This may be an over-simplification.[6]

The Parks model assumes vehicle values are fixed and will not be affected by the scrappage decision. This is reasonable given the random nature of the vehicle failure which leads to the need for the costly repair. There is no expectation that a repaired vehicle faces an increased probability of future failure.

The emissions tax, however, is recurring, even if the assessment amount varies from year to year. In this context, it is reasonable to assume that the vehicle owner would not discount future years' expected taxes completely when the expected useful life of the vehicle is greater than one year. Future years' expected taxes would affect the scrappage decision in one of two ways:

- The discounted expected future tax stream might be capitalized into the vehicle price (much as property taxes are capitalized into housing prices), causing the vehicle value to fall. The reduced vehicle value would increase the probability of scrappage.
- The vehicle value might remain unchanged because, unlike a house or land, the vehicle is mobile and could be offered for sale outside of the South Coast where the tax would not apply. In this case, the discounted expected lifetime tax would be compared with the unchanged vehicle value when the owner decides whether to keep the vehicle. Selling the vehicle out of the area would be an alternative to scrappage, effectively providing a positive salvage value to the scrappage calculation. Like the capitalization argument, this would also increase the probability of the vehicle being removed from the South Coast fleet. Because the emissions reduction in the South Coast would be expected to be the same whether the vehicle were scrapped or sold out of the region, this scenario can still be analysed in the same framework as a true scrappage decision.

It is possible that the expected useful life, future VMT and future emissions rates could be estimated probabilistically based on the vehicle's age and value, but this would ignore the likely dynamic adjustments in VMT and vehicle value over time. Such an analysis goes beyond the scope of the present study.

The methodology used in this study would still be applicable if future years' expected taxes are relevant to the scrappage decision. The probability of the discounted tax stream exceeding the vehicle value would increase but the trigger points or the decision-making process would not change. As such, this analysis will proceed with the simplifying assumption that only the current year's tax is relevant to the scrappage decision.

VMT is introduced into the model exogenously at its baseline level, based on the 1991 *RTECS* data. This implies that VMT is perfectly inelastic in the face of the emissions tax, a notion which the travel demand models developed in Chapter 3 argue strongly against. The baseline estimate, however, is useful for two reasons: it permits the estimation of the maximum potential scrappage for the South Coast fleet (which is useful for policy assessment); and it is free of any estimation errors from the VMT models. Re-estimating scrappage based on endogenously estimated VMT would permit a more complete simulation of the effects of the emissions tax. However, for many vehicles that are likely scrappage candidates, the emissions per mile are so high that even a substantial reduction in VMT would not alter the scrap-or-pay-the-tax decision.[7]

Each model year's scrappage probability is estimated using a decision-tree (expected value) framework. The model year's scrappage probability depends upon its own distributions of VMT, emissions and vehicle values. Scrappage probability for a vehicle of a given model year is measured as:

$$Prob(scrap) = \sum_{i=1}^{6} (VMT_i * VMTprob_i * \sum_{j=1}^{4} (Vehval_j * vehvalprob_j * taxprob))$$

where:

VMTprob = the probability that the vehicle's VMT will be in the given range;
Vehval = the vehicle's average book value (based on make, model, year);
Vehvalprob = the probability that the vehicle's book value will be in the given range;
Taxprob = the probability that the tax will exceed the vehicle's value, conditional upon the vehicle value and VMT range.

ESTIMATED MAXIMUM INDUCED SCRAPPAGE

The expected scrappage is calculated for each model year based on initial (pre-emissions tax) distributions of VMT, emissions and vehicle values by model year. Tables 5.3 to 5.8 show this calculation for 1976 to 1981 model year cars.

Scrappage probabilities for each model year represent the expected percentage of that year's vehicles facing emissions taxes in excess of vehicle value. More correctly, these probabilities show the maximum potential tax-induced scrappage, assuming future expected taxes are fully discounted. Some of the vehicles expected to be scrapped would, instead, undergo maintenance, be driven fewer miles or both, despite our assumption that optimal maintenance had been undertaken prior to the scrappage calculation. Some of the high-tax, low-value vehicles might have been scrapped anyway.

Scrappage may also fall short of the calculated expectation if the owner's subjective value of the vehicle exceeds the vehicle's book value. In this case, the owner would choose to pay the tax and keep the vehicle in use.

If the owner does not discount future taxes completely, induced scrappage would be greater than the rates indicated. For example, a $1000 vehicle would not be scrapped if its annual tax is $300. However, if the vehicle has a three-year remaining life and is expected to incur a $300 tax each year, the vehicle would be scrapped (or, more likely, offered for sale out of the region) if the future tax stream is discounted at a rate of 10 per cent because the future discounted tax of $746, added to the current year's tax of $300, would exceed the vehicle's value. Whether the vehicle is scrapped or removed from the region, the resulting emissions reduction is the same. If future years' taxes are considered, the resulting scrappage estimates could be considerably higher than those indicated in this analysis.

Despite these caveats, the baseline scrappage estimate is useful for policy analysis. Based on vehicle holdings by model year and the maximum percentage

Table 5.3 Expected induced scrappage of 1976 model year vehicles

VMT	Probability of VMT	Vehicle value ($)	Probability of vehicle value	Probability of tax > cut-off	Node value
		< 500	.084	.639	0.0273
< 5000	.508	500–1000	.535	.134	0.0365
		1001–1500	.016	.008	0.0001
		1501–2000	.011	.000	0.0000
		< 500	.084	.950	0.0293
5000–10000	.367	500–1000	.535	.639	0.1254
		1001–1500	.016	.303	0.0018
		1501–2000	.011	.134	0.0055
		< 500	.084	.975	0.0044
10001–15000	.054	500–1000	.535	.882	0.0255
		1001–1500	.016	.639	0.0006
		1501–2000	.011	.403	0.0024
		< 500	.084	1.000	0.0031
15001–20000	.037	500–1000	.535	.950	0.0188
		1001–1500	.016	.882	0.0005
		1501–2000	.011	.639	0.0026
		< 500	.084	1.000	0.0020
20001–25000	.024	500–1000	.535	.958	0.0123
		1001–1500	.016	.924	0.0004
		1501–2000	.011	.798	0.0021
		< 500	.084	1.000	0.0007
> 25000	.008	500–1000	.535	.975	0.0042
		1001–1500	.016	.950	0.0001
		1501–2000	.011	.882	0.0008

0.3064

→ expected induced scrappage = 30.64%

Taxing Automobile Emissions for Pollution Control

Table 5.4 Expected induced scrappage of 1977 model year vehicles

VMT	Probability of VMT	Vehicle value ($)	Probability of vehicle value	Probability of tax > cut-off	Node value
		< 500	.000	.696	0.0003
< 5000	.348	500–1000	.257	.187	0.0167
		1001–1500	.360	.042	0.0053
		1501–2000	.115	.000	0.0000
		< 500	.000	.965	0.0000
5000–10000	.493	500–1000	.257	.696	0.0882
		1001–1500	.360	.360	0.0640
		1501–2000	.115	.187	0.0106
		< 500	.000	1.000	0.0000
10001–15000	.085	500–1000	.257	.905	0.0198
		1001–1500	.360	.696	0.0213
		1501–2000	.115	.449	0.0044
		< 500	.000	1.000	0.0000
15001–20000	.029	500–1000	.257	.965	0.0072
		1001–1500	.360	.845	0.0088
		1501–2000	.115	.696	0.0023
		< 500	.000	1.000	0.0000
20001–25000	.014	500–1000	.257	.989	0.0036
		1001–1500	.360	.926	0.0047
		1501–2000	.115	.816	0.0013
		< 500	.000	1.000	0.0000
> 25000	.043	500–1000	.257	1.000	0.0111
		1001–1500	.360	.965	0.0149
		1501–2000	.115	.905	0.0045

0.2886

→ expected induced scrappage = 28.86%

Table 5.5 Expected induced scrappage of 1978 model year vehicles

VMT	Probability of VMT	Vehicle value ($)	Probability of vehicle value	Probability of tax > cut-off	Node value
		< 500	.000	.656	0.0000
< 5000	.321	500–1000	.083	.107	0.0029
		1001–1500	.532	.013	0.0022
		1501–2000	.011	.000	0.0000
		< 500	.000	.957	0.0000
5000–10000	.469	500–1000	.083	.656	0.0255
		1001–1500	.532	.309	0.0770
		1501–2000	.011	.107	0.0006
		< 500	.000	.997	0.0000
10001–15000	.124	500–1000	.083	.918	0.0095
		1001–1500	.532	.656	0.0432
		1501–2000	.011	.401	0.0005
		< 500	.000	1.000	0.0000
15001–20000	.062	500–1000	.083	.957	0.0049
		1001–1500	.532	.872	0.0288
		1501–2000	.011	.656	0.0004
		< 500	.000	1.000	0.0000
20001–25000	.012	500–1000	.083	.990	0.0010
		1001–1500	.532	.934	0.0060
		1501–2000	.011	.855	0.0001
		< 500	.000	1.000	0.0000
> 25000	.012	500–1000	.083	.997	0.0010
		1001–1500	.532	.957	0.0061
		1501–2000	.011	.918	0.0001

0.2098

→ expected induced scrappage = 20.98%

Table 5.6 Expected induced scrappage of 1979 model year vehicles

VMT	Probability of VMT	Vehicle value ($)	Probability of vehicle value	Probability of tax > cut-off	Node value
< 5000	.384	< 500	.000	.631	0.0000
		500–1000	.098	.110	0.0041
		1001–1500	.432	.009	0.0014
		1501–2000	.046	.000	0.0000
5000–10000	.424	< 500	.000	.935	0.0000
		500–1000	.098	.631	0.0262
		1001–1500	.432	.315	0.0576
		1501–2000	.046	.110	0.0021
10001–15000	.111	< 500	.000	.985	0.0000
		500–1000	.098	.841	0.0091
		1001–1500	.432	.631	0.0303
		1501–2000	.046	.397	0.0020
15001–20000	.030	< 500	.000	.996	0.0000
		500–1000	.098	.935	0.0027
		1001–1500	.432	.784	0.0102
		1501–2000	.046	.631	0.0009
20001–25000	.030	< 500	.000	1.000	0.0000
		500–1000	.098	.991	0.0029
		1001–1500	.432	.881	0.0114
		1501–2000	.046	.752	0.0010
> 25000	.000	< 500	.000	1.000	0.0000
		500–1000	.098	.985	0.0000
		1001–1500	.432	.935	0.0000
		1501–2000	.046	.841	0.0000

0.1622

→ expected induced scrappage = 16.22%

Table 5.7 Expected induced scrappage of 1980 model year vehicles

VMT	Probability of VMT	Vehicle value ($)	Probability of vehicle value	Probability of tax > cut-off	Node value
		< 500	.000	.497	0.0000
< 5000	.430	500–1000	.000	.077	0.0000
		1001–1500	.335	.011	0.0016
		1501–2000	.231	.000	0.0000
		< 500	.000	.835	0.0000
5000–10000	.329	500–1000	.000	.497	0.0000
		1001–1500	.335	.192	0.0212
		1501–2000	.231	.077	0.0058
		< 500	.000	.967	0.0000
10001–15000	.165	500–1000	.000	.709	0.0000
		1001–1500	.335	.497	0.0275
		1501–2000	.231	.288	0.0110
		< 500	.000	.989	0.0000
15001–20000	.063	500–1000	.000	.835	0.0000
		1001–1500	.335	.654	0.0138
		1501–2000	.231	.497	0.0072
		< 500	.000	1.000	0.0000
20001–25000	.013	500–1000	.000	.931	0.0000
		1001–1500	.335	.766	0.0033
		1501–2000	.231	.621	0.0019
		< 500	.000	1.000	0.0000
> 25000	.000	500–1000	.000	.967	0.0000
		1001–1500	.335	.835	0.0000
		1501–2000	.231	.709	0.0000

0.0933

→ expected induced scrappage = 9.33%

Table 5.8 Expected induced scrappage of 1981 model year vehicles

VMT	Probability of VMT	Vehicle value ($)	Probability of vehicle value	Probability of tax > cut-off	Node value
		< 500	.000	.114	0.0000
< 5000	.181	500–1000	.000	.019	0.0000
		1001–1500	.000	.000	0.0000
		1501–2000	.096	.000	0.0000
		< 500	.000	.452	0.0000
5000–10000	.500	500–1000	.000	.114	0.0000
		1001–1500	.000	.032	0.0000
		1501–2000	.096	.019	0.0009
		< 500	.000	.646	0.0000
10001–15000	.236	500–1000	.000	.280	0.0000
		1001–1500	.000	.114	0.0000
		1501–2000	.096	.056	0.0013
		< 500	.000	.817	0.0000
15001–20000	.042	500–1000	.000	.452	0.0000
		1001–1500	.000	.235	0.0000
		1501–2000	.096	.114	0.0005
		< 500	.000	.905	0.0000
20001–25000	.014	500–1000	.000	.566	0.0000
		1001–1500	.000	.349	0.0000
		1501–2000	.096	.193	0.0003
		< 500	.000	.950	0.0000
> 25000	.028	500–1000	.000	.646	0.0000
		1001–1500	.000	.452	0.0000
		1501–2000	.096	.280	0.0008

0.0036

→ expected induced scrappage = 0.36%

of each year's holdings that would be scrapped, it is possible to estimate the maximum number of vehicles that would be scrapped by model year. From these figures, the *Mobile Source Emission Reduction Credits Guidelines* (1993) established by CARB for the RECLAIM emissions trading programme can be used to estimate the emissions reduction that would result from the scrappage.[8] Table 5.9 shows the emissions credits for scrapping a vehicle of a given vintage.

Table 5.9   Net emissions reduction credit per vehicle scrapped

Model year	HC (lbs/year)	CO (lbs/year)	NO_x (lbs/year)
Pre-1972	110	600	30
1972–74	93	380	30
1975–81	32	320	25

Source: CARB (1993), Table 5, p. 29.

Table 5.10 shows that a maximum of 21.5 per cent of the pre-1982 fleet that existed in 1990 would have been induced by the emissions tax to be scrapped, compared with a natural scrappage rate of 12–15 per cent of older vehicles.[9]

Table 5.10   Maximum expected scrappage and resulting emissions reduction

Model year	Vehicles in fleet	% of fleet scrapped	Vehicles scrapped	HC saved (tons/yr)	CO saved (tons/yr)	NO_x saved (tons/yr)
1981	232 030	0.36	845	13.52	135.20	10.56
1980	231 731	9.33	21 621	345.93	3459.28	270.26
1979	233 476	16.22	37 870	605.92	6059.17	473.37
1978	225 941	20.98	47 402	758.44	7584.39	592.53
1977	181 573	28.86	52 402	838.43	8384.31	655.02
1976	126 575	30.64	38 783	620.52	6205.21	484.78
pre-1976	723 166	30.64	221 577	10 443.54	54 119.97	3253.70
Total	1 954 492	21.51	420 500	13 626.30	85 947.53	5740.22
TPD*				37.33	235.47	15.73

Note: * TPD = tons per day assuming 365 days per year.

It is not possible to determine which of the vehicles scrapped in response to the emissions tax would have been part of the natural scrappage. In all likelihood, there is considerable overlap. The emissions savings from vehicle scrappage represent 5.98, 9.28 and 5 per cent of the 1994 targeted reductions of ROG, CO and NO_x for the South Coast, as reported in the 1991 *Air Quality Management Plan* (SCAQMD, 1991).

This scrappage analysis could be extended to consider the role of the Mobile Source Emission Reductions Credit Programmes which provide a financial incentive for vehicle owners to scrap their low-value cars. The potential scrappage candidates identified in the above analysis are prime candidates to sell their high-emitting vehicles into this programme; any payment received above the scrap value would be an added benefit to the vehicle owner looking to unload his or her tax liability. These would be self-selected high emitters, ensuring that the emissions trading credits do not allow more stationary source emissions than what was removed from the mobile source.

SCRAPPAGE CONDITIONAL ON REDUCED VMT

The scrappage analysis, conditional on *ex ante* VMT, emissions and vehicle value, assumes the vehicle owner will not adjust VMT in response to the emissions tax. TIERS results suggest otherwise, namely that VMT will be reduced substantially for high-emitting vehicles. It is likely that inexpensive maintenance and reduced VMT would be the first responses to the tax. If the resulting VMT and emissions still produce an annual tax which exceeds the vehicle value, the vehicle would be expected to be scrapped.

Tables 5.11 to 5.17 show the effect of reduced VMT on expected scrappage. The VMT probabilities have been re-estimated using the TIERS average VMT reduction of 41 per cent. This approach assumes the tax would be levied on the miles that remained after the VMT adjustment occurs.

Scrappage conditional on vehicle values and reduced VMT is estimated to be only 12.4 per cent of the fleet, down from 21.5 per cent when the unadjusted, pre-emissions tax, VMT was used. The 12.4 per cent is well within the natural scrappage rate for older vehicles. Again, it is not possible to determine which of these vehicles induced to scrappage by the emissions tax would have been scrapped for other reasons in the absence of the tax.

Table 5.11 Adjusted expected scrappage of 1976 model year vehicles

VMT	Probability of VMT	Vehicle value ($)	Probability of vehicle value	Probability of tax > cut-off	Node value
		< 500	.084	.639	0.0443
< 5000	.825	500–1000	.535	.134	0.0591
		1001–1500	.016	.008	0.0001
		1501–2000	.011	.000	0.0000
		< 500	.084	.950	0.0096
5000–10000	.120	500–1000	.535	.639	0.0410
		1001–1500	.016	.303	0.0006
		1501–2000	.011	.134	0.0002
		< 500	.084	.975	0.0039
10001–15000	.048	500–1000	.535	.882	0.0226
		1001–1500	.016	.639	0.0005
		1501–2000	.011	.403	0.0002
		< 500	.084	1.000	0.0006
15001–20000	.007	500–1000	.535	.950	0.0036
		1001–1500	.016	.882	0.0001
		1501–2000	.011	.639	0.0000
		< 500	.084	1.000	0.0000
20001–25000	.000	500–1000	.535	.958	0.0000
		1001–1500	.016	.924	0.0000
		1501–2000	.011	.798	0.0000
		< 500	.084	1.000	0.0000
> 25000	.000	500–1000	.535	.975	0.0000
		1001–1500	.016	.950	0.0000
		1501–2000	.011	.882	0.0000

|||||| 0.1865 |

→ adjusted expected scrappage = 18.65%

Table 5.12 Adjusted expected scrappage of 1977 model year vehicles

VMT	Probability of VMT	Vehicle value ($)	Probability of vehicle value	Probability of tax > cut-off	Node value
		< 500	.000	.696	0.0000
< 5000	.681	500–1000	.257	.187	0.0327
		1001–1500	.360	.042	0.0103
		1501–2000	.115	.000	0.0000
		< 500	.000	.965	0.0000
5000–10000	.246	500–1000	.257	.696	0.0440
		1001–1500	.360	.360	0.0319
		1501–2000	.115	.187	0.0053
		< 500	.000	1.000	0.0000
10001–15000	.029	500–1000	.257	.905	0.0067
		1001–1500	.360	.696	0.0073
		1501–2000	.115	.449	0.0015
		< 500	.000	1.000	0.0000
15001–20000	.043	500–1000	.257	.965	0.0107
		1001–1500	.360	.845	0.0131
		1501–2000	.115	.696	0.0034
		< 500	.000	1.000	0.0000
20001–25000	.000	500–1000	.257	.989	0.0000
		1001–1500	.360	.926	0.0000
		1501–2000	.115	.816	0.0000
		< 500	.000	1.000	0.0000
> 25000	.000	500–1000	.257	1.000	0.0000
		1001–1500	.360	.965	0.0000
		1501–2000	.115	.905	0.0000

0.1669

→ adjusted expected scrappage = 16.69%

Table 5.13 Adjusted expected scrappage of 1978 model year vehicles

VMT	Probability of VMT	Vehicle value ($)	Probability of vehicle value	Probability of tax > cut-off	Node value
< 5000	.667	< 500	.000	.656	0.0000
		500–1000	.083	.107	0.0059
		1001–1500	.532	.013	0.0046
		1501–2000	.011	.000	0.0000
5000–10000	.259	< 500	.000	.957	0.0000
		500–1000	.083	.656	0.0141
		1001–1500	.532	.309	0.0426
		1501–2000	.011	.107	0.0003
10001–15000	.062	< 500	.000	.997	0.0000
		500–1000	.083	.918	0.0047
		1001–1500	.532	.656	0.0216
		1501–2000	.011	.401	0.0003
15001–20000	.012	< 500	.000	1.000	0.0000
		500–1000	.083	.957	0.0010
		1001–1500	.532	.872	0.0056
		1501–2000	.011	.656	0.0001
20001–25000	.000	< 500	.000	1.000	0.0000
		500–1000	.083	.990	0.0000
		1001–1500	.532	.934	0.0000
		1501–2000	.011	.855	0.0000
> 25000	.000	< 500	.000	1.000	0.0000
		500–1000	.083	.997	0.0000
		1001–1500	.532	.957	0.0000
		1501–2000	.011	.918	0.0000

0.1008

→ adjusted expected scrappage = 10.08%

Table 5.14 Adjusted expected scrappage of 1979 model year vehicles

VMT	Probability of VMT	Vehicle value ($)	Probability of vehicle value	Probability of tax > cut-off	Node value
		< 500	.000	.631	0.0000
< 5000	.657	500–1000	.098	.110	0.0071
		1001–1500	.432	.009	0.0026
		1501–2000	.046	.000	0.0000
		< 500	.000	.935	0.0000
5000–10000	.293	500–1000	.098	.631	0.0181
		1001–1500	.432	.315	0.0399
		1501–2000	.046	.110	0.0015
		< 500	.000	.985	0.0000
10001–15000	.051	500–1000	.098	.841	0.0042
		1001–1500	.432	.631	0.0139
		1501–2000	.046	.397	0.0009
		< 500	.000	.996	0.0000
15001–20000	.000	500–1000	.098	.935	0.0000
		1001–1500	.432	.784	0.0000
		1501–2000	.046	.631	0.0000
		< 500	.000	1.000	0.0000
20001–25000	.000	500–1000	.098	.991	0.0000
		1001–1500	.432	.881	0.0000
		1501–2000	.046	.752	0.0000
		< 500	.000	1.000	0.0000
> 25000	.000	500–1000	.098	.985	0.0000
		1001–1500	.432	.935	0.0000
		1501–2000	.046	.841	0.0000

0.0881

→ adjusted expected scrappage = 8.81%

Table 5.15 Adjusted expected scrappage of 1980 model year vehicles

VMT	Probability of VMT	Vehicle value ($)	Probability of vehicle value	Probability of tax > cut-off	Node value
		< 500	.000	.497	0.0000
< 5000	.633	500–1000	.000	.077	0.0000
		1001–1500	.335	.011	0.0023
		1501–2000	.231	.000	0.0000
		< 500	.000	.835	0.0000
5000–10000	.329	500–1000	.000	.497	0.0000
		1001–1500	.335	.192	0.0212
		1501–2000	.231	.077	0.0059
		< 500	.000	.967	0.0000
10001–15000	.038	500–1000	.000	.709	0.0000
		1001–1500	.335	.497	0.0063
		1501–2000	.231	.288	0.0025
		< 500	.000	.989	0.0000
15001–20000	.000	500–1000	.000	.835	0.0000
		1001–1500	.335	.654	0.0000
		1501–2000	.231	.497	0.0000
		< 500	.000	1.000	0.0000
20001–25000	.000	500–1000	.000	.931	0.0000
		1001–1500	.335	.766	0.0000
		1501–2000	.231	.621	0.0000
		< 500	.000	1.000	0.0000
> 25000	.000	500–1000	.000	.967	0.0000
		1001–1500	.335	.835	0.0000
		1501–2000	.231	.709	0.0000

0.0382

→ adjusted expected scrappage = 3.82%

Table 5.16 Adjusted expected scrappage of 1981 model year vehicles

VMT	Probability of VMT	Vehicle value ($)	Probability of vehicle value	Probability of tax > cut-off	Node value
< 5000	.528	< 500	.000	.114	0.0000
		500–1000	.000	.019	0.0000
		1001–1500	.000	.000	0.0000
		1501–2000	.096	.000	0.0000
5000–10000	.417	< 500	.000	.452	0.0000
		500–1000	.000	.114	0.0000
		1001–1500	.000	.032	0.0000
		1501–2000	.096	.019	0.0008
10001–15000	.028	< 500	.000	.646	0.0000
		500–1000	.000	.280	0.0000
		1001–1500	.000	.114	0.0000
		1501–2000	.096	.056	0.0002
15001–20000	.014	< 500	.000	.817	0.0000
		500–1000	.000	.452	0.0000
		1001–1500	.000	.235	0.0000
		1501–2000	.096	.114	0.0002
20001–25000	.014	< 500	.000	.905	0.0000
		500–1000	.000	.566	0.0000
		1001–1500	.000	.349	0.0000
		1501–2000	.096	.193	0.0003
> 25000	.000	< 500	.000	.950	0.0000
		500–1000	.000	.646	0.0000
		1001–1500	.000	.452	0.0000
		1501–2000	.096	.280	0.0000

0.0015

→ expected induced scrappage = 0.15%

Table 5.17 Adjusted expected scrappage and resulting emissions reduction

Model year	Vehicles in fleet	% of fleet scrapped	Vehicles scrapped	HC saved (tons/yr)	CO saved (tons/yr)	NO$_x$ saved (tons/yr)
1981	232 030	0.15	345	5.57	55.68	4.35
1980	231 731	3.82	8852	141.63	1416.32	110.65
1979	233 476	8.81	20 569	329.10	3291.04	257.11
1978	225 941	10.08	22 775	364.40	3644.00	284.69
1977	181 573	16.69	30 304	484.86	4848.64	378.80
1976	126 575	18.65	23 606	377.70	3776.96	295.08
pre-1976	723 166	18.65	134 871	6 356.83	32 941.96	1980.48
Total	1 954 492	12.35	241 325	8 060.09	49 974.60	3311.16
TPD*				22.08	136.92	9.07

Note: * TPD = tons per day assuming 365 days per year.

SCRAPPAGE SUMMARY AND IMPLICATIONS

This chapter has extended the Parks model of vehicle scrappage to examine the effect of the emissions tax on the decision to keep or scrap each vehicle. The baseline analysis assumed the scrappage decision would be made before any VMT adjustment. The adjusted scrappage analysis assumed scrappage occurred after VMT reductions. Both analyses showed that the emissions tax would provide an incentive for scrapping high-emitting vehicles. This has important implications for vehicle buy-back programmes, which have been only somewhat successful at identifying high emitters.[10]

A vehicle buy-back programme would increase the number of vehicles scrapped. The scrappage decision would then be based on whether the vehicle value exceeded the sum of the tax plus the buy-back price. The buy-back price would be the opportunity cost of keeping the vehicle.

Although VMT declines have been shown to lead to greater emissions reduction than induced scrappage, removing high-emitting vehicles from the fleet creates a permanent emissions saving. VMT reductions could be reversed in the future, especially if fuel cost declines. This makes scrappage an especially important component of the response to the emissions tax.

NOTES

1. Durability characteristics are imbedded in the vehicle at the time it is produced, making the vehicle more or less prone to breakdowns.
2. The implicit assumption is that the emissions tax would be waived if the vehicle were scrapped.
3. Emissions are treated as a composite unit in this specification, although the mobile source emissions tax would measure emissions of HC, CO and NO_x and apply the appropriate tax rate to each pollutant.
4. The vehicle owner expects, *a priori*, that maintenance will reduce emissions. Sommerville et al.'s (1987) results indicate that maintenance may increase emissions of one or even two pollutants. There is, in particular, an inverse relationship between CO and NO_x emissions.
5. Such as the NADA or Edmunds used car price guides.
6. Hensher (1987) discounted future car values by 10 per cent per year in his scrappage model, based on a three-year remaining useful life.
7. For example, 11.3 per cent of vehicles scrapped by the Illinois EPA in its pilot scrappage programme would have incurred tax rates in excess of $0.50 per mile. All of these vehicles were willingly sold by their owners for less than $1000, indicating low perceived value. Their VMT would have to drop below 2000 miles to incur taxes under $1000. Such low-use, low-value vehicles are prime candidates for scrappage.
8. CARB guidelines provide emissions reduction credits for vehicle scrappage based on model year. The credits are *net* reductions because they already reflect the expected additional emissions from an average replacement vehicle.
9. Natural scrappage rates are calculated based on the year to year change in the number of vehicles registered by model year. Data sources include annual publications of *Motor Vehicle Facts and Figures* by the American Automobile Manufacturers Association and various editions of the *Transportation Energy Data Book* by the Oak Ridge National Laboratory (Davis, 1994).
10. Even programmes which identified and solicited participation from vehicles that had high emissions rates at testing facilities, such as the Cash for Clunkers programme run by the Illinois EPA, did not know the pattern of vehicle usage. Among vehicles scrapped in the Illinois programme whose odometers were believed to be reliable, annual VMT ranged from 133 to 30465 miles. Annualized emissions from the dirtiest car in the Illinois programme exceeded one metric tonne of HC, while the cleanest emitted only 365 grams. The emissions tax alone would not have encouraged the latter to be scrapped, but the $19616 tax (based on HC alone!) would have been a strong motivator for the dirty vehicle which was scrapped for $902 (Illinois EPA, 1993, Appendix, p. 3).

6. Summary and Conclusions

California's South Coast Air Basin has the nation's highest concentrations of ground-level ozone and is required to formulate and enact policies to reduce the emissions that lead to ozone formation. Least cost emissions reduction defines an efficient solution to the need to abate, but there is no guarantee that a particular policy designed to reduce emissions will do so at the least possible cost.

An emissions tax is theoretically known to reduce emissions in a cost-effective manner by providing the emitter with an incentive to abate up to the point where the marginal abatement cost equals the unit tax. Such a tax has been discussed for mobile sources, and either embraced in theory (but without analysis) or simply dismissed as too cumbersome given the number of vehicles. Neither approach is satisfactory.

Given the strong theoretical foundation for an emissions tax and the lack of empirical work in this area,[1] this study has attempted to develop and present an empirical study of the expected effects of instituting an emissions tax on household motor vehicles in California's South Coast Air Basin. This study focused on five topics:

1. Developing appropriate tax rates for each of the three primary mobile source pollutants – HCs, CO and NO_x – and a practical means for estimating each vehicle's annual emissions.
2. Developing models of household travel behaviour (VMT) to assess the response to the tax.
3. Integrating data on vehicle ownership, operating cost, emissions and demand elasticities into a simulation model (TIERS) designed to assess the changes in area-wide travel and emissions.
4. Developing and estimating a model of vehicle scrappage.
5. Assessing the revenue and equity aspects of the tax.

The lack of a single dataset containing information on household characteristics, vehicle ownership, trips, VMT, vehicle operating costs and emissions led to the use of a combination of national and regional measures and estimates to perform the empirical analyses. Such data handling adds a risk of error to any empirical study. However, a careful analysis of the data revealed strong similarities between California and Midwestern vehicle emissions and between the CalTrans and CARB distributions of vehicle ownership in the South Coast. While not ideal, the data appear to be adequate for this study.

The results of the simulations performed with the TIERS model show the potential importance of a mobile source emissions tax in the South Coast region. In 1991, mobile sources (including heavy trucks and buses) accounted for 44 per cent of the area's ROG emissions, 87 per cent of CO and 55 per cent of NO_x (SCAQMD, 1991). If mobile source emissions are not reduced substantially, the South Coast will fail to attain the NAAQS by 2010, as required by the 1990 CAAA (SCAQMD, 1991).

While the emissions tax under discussion would affect only household vehicles, this study has shown that the potential emissions reduction is considerable, amounting to one-third to one-half of the region's targeted 1994 reductions in ROG and NO_x and one-half to three-quarters of the CO target. These are sizeable impacts in an area which has already invested heavily in technology and regulations designed to achieve compliance with the NAAQS.

The TIERS model has shown that the expected emissions reduction is greater than the expected reduction in VMT, regardless of the magnitude of the demand elasticities used. This is also a significant result. The goal of the tax is to reduce emissions, not to reduce travel. Pricing travel to reflect the social cost of emissions leads to a more efficient allocation of travel because the costliest miles (those with the highest emissions rates) are reduced first. Similarly, the vehicles induced to scrappage are not necessarily the oldest, or the dirtiest, but the ones whose value is less than the cost of their annual emissions. This is also efficiency enhancing.

The questions of tax equity were addressed to determine if the incidence of the tax would be borne disproportionately by lower-income households. Simulations run over separate income quintiles showed that the tax for one-vehicle households would be about 15 per cent higher for the highest income quintile compared with the lowest income quintile ($134 vs $117) and about 22 per cent higher for multi-vehicle households ($390 vs $311). However, lower-income households would reduce travel by a greater percentage (42 vs 35 per cent) than upper-income households. As income rises, households tend to own more vehicles, allowing greater substitution which may offset some of the effects of the tax, and more newer vehicles which, on average, face lower emissions taxes. Not surprisingly, the emissions tax would be regressive.

In the absence of evasion or vehicle tampering, the tax would be expected to raise nearly $1.2 billion in revenue. Although the SCAQMD has studied potential emissions-reducing expenditures for revenue raised through the pricing of transportation emissions,[2] public finance theory argues in favour of reducing other taxes in the region which are more distorting in order to improve the overall efficiency of the tax system. This study has shown that a reduction in the gasoline tax could be efficiency enhancing, but there are unresolved jurisdictional questions in this regard.[3] A reduction in the annual motor vehicle registration and licensing fees could be revenue-neutral but it would not improve the efficiency of the tax system.

FURTHER WORK

This study should be viewed as a starting point for work in a number of important areas. It has raised issues about the data needed to study travel behaviour and vehicle ownership, the nature of household travel demand, the scrappage decision, the costs and benefits of vehicle maintenance, vehicle tampering and the enforcement of regulations designed to limit mobile source emissions. Each issue can be developed further.

This study has demonstrated that it is possible to study the effects of an emissions tax levied on individual motor vehicles by integrating data on vehicle ownership, characteristics and use from several diverse sources. It would be far better to develop a single, comprehensive household travel survey which includes individual vehicle VMT data by trip, vehicle emissions and characteristics of the individual drivers. This would provide a sounder basis for analysing the effects of pricing on vehicle usage. Panel data, such as that developed by Hensher et al. (1992) in Australia, would provide the basis for a dynamic study of vehicle ownership as well as allow for the development of a long-run VMT price elasticity. As the vehicle fleet changes over time, the resulting travel and emissions would also be altered. The TIERS model is not suited to study these dynamics.

Running TIERS under assumptions of different price elasticities led to an important unanswered question. Is the decision-making unit the household or the individual driver when VMT decisions are made? If it is the latter, the use of household-based VMT models may be of less use than models based on individual drivers, as Golub et al. (1994) have begun to explore. This is an important area for further study.

The scrappage study has shown that the tax would provide an effective mechanism for inducing only those low-value vehicles that are high emitters to be scrapped. RECLAIM encourages the purchase of older vehicles to generate emissions credits but has no effective mechanism for separating high-emitting vehicles from low emitters. A model of emissions trading incorporating the expected supply of high-emitting vehicles facing a high tax would enable policy-makers to assess the full potential of RECLAIM.

This study has focused on VMT and exhaust emissions but, as Chapter 2 discusses, cold start and hot soak emissions can account for a significant percentage of total trip emissions, especially for older vehicles operating in cold weather. A more in-depth study of emissions should consider trips as well as

VMT. (The emissions tax presented in this study could, in fact, have the perverse consequence of increasing vehicle emissions by encouraging more short trips to reduce taxable VMT.) A trip generation model, accounting for trip cost, would be a useful addition to TIERS.

This study also suggests that a comprehensive analysis of evasion potential and the costs of enforcement and compliance would be necessary before an emissions tax could be considered as a serious policy instrument in the South Coast. The analyses presented have implicitly or explicitly assumed full compliance with the emissions tax. However, widespread tampering with vehicle odometers and/or emissions control systems could undermine the effectiveness of the tax.

As Lawson (1993) indicates, a periodic, anticipated vehicle inspection allows vehicle owners to do whatever is expedient to pass the test, but does not encourage them to maintain that performance level. Temporary adjustment of the emissions control system is often a simple, low-cost (and often undetectable) means of achieving an acceptably low emissions reading at inspection time. Yet the current pass/fail inspection system provides less incentive to present the vehicle with minimized emissions than would occur when the vehicle owner would have to pay a tax for the emissions.

Random testing of on-road vehicles, either by remote sensing or by setting up portable field-testing equipment would likely be needed to ensure year-round emissions savings. Remote sensing, in which the exhaust stream of a passing vehicle is analysed and the licence plate is read and recorded, has been field tested in many independent studies. The technology captures readings of HC and CO. While there are valid questions regarding the reliability of a single remote reading compared with the test conditions of the I/M 240, most studies have shown that remote sensing is generally able to identify high emitters (Cadle et al., 1994). Since the sensors would not be used to assess the emissions tax, but as a deterrent against allowing the vehicle's emissions to increase between scheduled inspections, their level of accuracy would not pose a serious problem. Vehicles identified by the sensors as high emitters could be required to undergo a roadside inspection to reassess the emissions tax. Penalties for tampering could also be imposed at that time.

Odometer tampering is a more difficult issue, especially for older vehicles which may have been purchased with faulty equipment. Hub odometers, such as those used by commercial vehicles, may provide a solution to this problem. Harsher penalties for tampering may provide the strongest disincentive to potential abusers. Again, this issue warrants further study.

Additional costs will have to be incurred to ensure that vehicles do not increase their emissions following the I/M 240 test which assesses the tax. A comprehensive study, outlining the personnel and equipment needed to attain a targeted level of compliance, along with the costs, is suggested. Without an enforcement effort to parallel the tax assessment, much of the potential emissions reduction could be lost.

Several important issues have not been addressed, but would be worthwhile extensions of the work presented. These include: the possibility of taxing commercial vehicle emissions; effects on residential and commercial location decisions; effects of reduced VMT on roadway congestion and travel speed; mode choice and substitution; and the political feasibility of an emissions tax. I shall discuss each briefly.

Although this study has been limited to taxing the emissions of household vehicles, it might be useful to consider a similar tax on commercial vehicles. Heavy duty vehicles, especially those fuelled by diesel, are significant emitters of NO_x and small particulate matter (SCAQMD, 1991). Taxing these vehicles' emissions and reducing the tax on diesel fuel would be a logical extension to a tax on household vehicles' emissions.

Roadway congestion is a serious transportation externality in the South Coast, increasing the time and fuel costs of travel. Low vehicle speed also increases vehicle emissions, as discussed in Chapter 2. If the emissions tax reduces VMT, it could also reduce congestion. The extent of this effect would depend on the timing of the VMT which is reduced. Since the tax provides no incentive to travel at certain times of the day, and does not distinguish between commuting and non-work travel, it is possible that commuting miles would continue to be maintained and concentrated at the same time of day. This would not reduce congestion. However, given that commuting accounts for only 25 per cent of the trips in the South Coast (Cambridge Systematics, Inc., 1993), non-work travel would also have to be reduced to accomplish the VMT reductions indicated by TIERS. The net result is that congestion would likely be reduced in the short run, but the issue merits further study.

In the long run, however, free-flowing traffic often encourages increased development along the uncongested roadway. This, in turn, can re-establish congestion. Congestion should be studied in the context of long-run land-use decisions which go well beyond the scope of this study. The land-use questions are critically important, however, because the challenge in the South Coast is not merely to meet the NAAQS by 2010 but to continue to meet the standards as population and development increase over time.

This study has presented strong evidence that the private vehicle is the preferred mode of travel. The sharp reductions in VMT projected by TIERS would be expected to be accompanied by an increased demand for travel by alternative modes, including car-pooling. A mode choice model focused on car-pooling, rather than transit, as the alternative to single-occupancy driving, would be useful to study this effect.

Finally, and reluctantly, the question of political feasibility must be addressed. In an era of conflicting priorities, when 'no new taxes' and 'less government regulation' are constraints binding on most policy-makers' minds, it seems a mobile source emissions tax may be more of a rigorous academic exercise than a tool for addressing a serious problem. California has already successfully challenged USEPA's requirement that all private vehicles undergo an annual I/M 240 emissions inspection at a centralized test facility.[4] Can the SCAQMD seriously recommend an emissions tax, along with increased enforcement, in this setting? The real question may be, is the South Coast serious about its goal of 'clean air, at the lowest possible cost'? If so, it may be missing an opportunity if it dismisses a mobile source emissions tax from serious consideration at this time.

This study provides a rationale, framework and methodology for studying the use of a mobile source emissions tax to reduce emissions in the South Coast Air Basin. Better data and more robust travel demand models, combined with dynamic models of vehicle ownership, a better understanding of vehicle maintenance and a thorough analysis of the institutions charged with implementing and administering the tax and monitoring compliance, would go a long way to improving the analysis presented herein. The conclusion that a mobile source emissions tax has the potential to reduce emissions significantly in a cost-effective, revenue-neutral manner is an exciting and important first step for further policy analysis.

NOTES

1. Cambridge Systematics, Inc. (1993) attempted to analyse a mobile source emissions tax but, in fact, their study analysed a VMT tax. Cameron (1991) also analysed a one cent per mile tax. Given the wide variability of emissions per mile, these studies should not be considered empirical analyses of a mobile source emissions tax.
2. Cambridge Systematics, Inc. (1993) evaluated 13 potential revenue reinvestment projects, each designed to reduce mobile source emissions in the South Coast. Among such projects were improved transit, employer-sponsored car-pooling and buying older vehicles for scrappage.
3. The emissions tax would affect only four counties but the gasoline tax is levied state-wide. The gasoline tax revenue raised within the South Coast counties is insufficient to offset the projected emissions tax revenue.
4. In 1994, USEPA agreed to allow California to continue its biannual Smog Check at private garages. Vehicles that failed the test would be required to be re-tested at a centralized facility.

Appendix A: Federal Requirements for Non-attainment Areas

By 1996, all areas except those classified as marginal for ozone non-attainment must achieve a 15 per cent reduction from all sources – stationary, area and mobile – that is, demonstrate 'reasonable further progress'. In addition, the South Coast region must demonstrate annual reductions averaging 3 per cent in successive years.

Table A.1 NAAQS and ozone non-attainment designation

Designation	One-hour concentration (ppm)	Target for attainment
Marginal	0.121–0.138	1993
Moderate	0.138–0.160	1996
Serious	0.160–0.180	1999
Severe 1	0.180–0.190	2005
Severe 2	0.190–0.280	2007
Extreme	> 0.280	2010

Table A.2 NAAQS and CO non-attainment designation

Designation	Eight-hour concentration (ppm)	Target for attainment
Moderate	9.1–16.4	1996
Serious	> 16.4	2000

Table A.3 Requirements for ozone non-attainment areas

Designation	Actions required	Due
Marginal	– Submit emissions inventory of all HC sources in 1990. – Revise every three years until attainment.	15/11/92
Moderate	– Revise State Implementation Plan (SIP) to show control strategies to reduce HC baseline emissions 15% over the period 1990–96. – The 15% reduction must accommodate VMT growth resulting from population growth. – EPA regulations promulgated by 1/1/90 regarding vehicle emissions control systems, fuel volatility or measures to correct SIP or I/M deficiencies cannot count towards the 15% reduction.	15/11/93
Serious, severe and extreme	– Demonstrate a reduction of 3% per year on average after 1996 until attainment. – Demonstrate air quality attainment using an EPA-approved photochemical dispersion model or other analytical method. – Demonstrate that aggregate VMT, vehicle emissions and congestion levels are consistent with projections used to demonstrate attainment. – If current levels exceed projected levels, the state must submit a SIP revision within 18 months that includes strategies to reduce emissions to their projected levels.	15/11/94

Note: Requirements are cumulative.

Table A.4 Specific requirements for ozone reduction

Designation	Actions required
Marginal	– Implement current SIP commitments; correct deficiencies.
	– Basic I/M should meet the stricter of EPA or SIP guidance if such a programme were required before enactment of the CAAA.
Moderate	– Must include basic I/M in the SIP.
	– Submit Stage II vapour recovery programme for stations selling over 10000 gallons/month (50000 gallons for small independent businesses).
	– Contingency provisions in the form of TCMs or other measures must be provided in the 1993 SIP. These measures will take effect without further action by the state or EPA at any point the state fails to meet the 15% emissions reduction targets required by 1996, fails to attain the NAAQS target date or, for regions designated serious and above, fails to meet the 3% annual emissions reductions required after 1996.
Serious	– Enhanced I/M programme must be submitted by 15/11/92.
	– Clean-Fuel Fleet Programme for centrally fuelled fleets of ten or more vehicles in regions with 1980 population exceeding 250000. SIP must ensure effectiveness of programme.
Severe 1 and Severe 2	– Submit specific strategies and measures by 15/11/92 to offset emissions from growth in VMT or number of trips.
	– Submit SIP revision by 15/11/92 detailing trip reduction programme for employers with 100 or more employees. The programme must increase average vehicle occupancy by at least 25% above the area's average. Employer compliance plans due two years after SIP submittal. Plans should convincingly demonstrate compliance four years after SIP submittal.
	– Reformulated gasoline required by 1995.
Extreme	– May submit additional measures to reduce the use of high-polluting heavy duty vehicles during peak traffic hours.

Note: Requirements are cumulative.

Appendix B: Section 108(f)(1) Transportation Control Measures

1. Programmes for improved public transit.
2. Restrictions of certain roads or lanes to, or construction of such roads or lanes for use by, passenger buses or high-occupancy vehicles.
3. Employer-based transportation management plans, including incentives.
4. Trip reduction ordinances.
5. Traffic flow improvement programmes that achieve emissions reductions.
6. Fringe and transportation corridor parking facilities serving multiple-occupancy vehicle programmes or transit services.
7. Programmes to limit or restrict vehicle use in downtown areas or other areas of emissions concentration, particularly during periods of peak use.
8. Programmes for the provision of all forms of high-occupancy, shared-ride services.
9. Programmes to limit portions of road surfaces or certain sections of the metropolitan area to the use of non-motorized vehicles or pedestrian use, both as to time and place.
10. Programmes for secure bicycle storage facilities and other facilities, including bicycle lanes, for the convenience and protection of bicyclists, in both public and private areas.
11. Programmes to control extended idling of vehicles.
12. Programmes to reduce motor vehicle emissions consistent with Title II, which are caused by extreme cold start conditions.
13. Employer-sponsored programmes to permit flexible work schedules.
14. Programmes and ordinances to facilitate non-automobile travel, provision and utilization of mass transit and generally to reduce the need for single-occupant vehicle travel, as part of transportation planning and development efforts of a locality, including programmes and ordinances applicable to new shopping centres, special events and other centres of vehicle activity.
15. Programmes for new construction and major reconstruction of paths, tracks or areas solely for use by pedestrian or other non-motorized means of transportation when economically feasible and in the public interest.
16. Programmes to encourage the voluntary removal from use and the marketplace of pre-1980 model year light duty vehicles and pre-1980 light duty trucks.

Appendix C: Data Used in this Study

Data required for analysing the expected effects of a mobile source emissions tax have been assembled from the following sources:

1. Vehicle values in the South Coast by model year: pre-1977 to 1992 (source: CEC's own database).
2. The number of passenger cars in the South Coast fleet in 1990 by model year (source: CARB's own database).
3. Individual vehicle emissions of HC, CO and NO_x for 1976 to 1992 model year cars (source: USEPA Hammond, Indiana I/M 240 testing facility lane data).
4. US Department of Energy, 1991 *Residential Transportation Energy Consumption Survey* (*RTECS*) provided data on household demographics, vehicle holdings, operating costs and VMT.
5. Tax rates of $15000/ton of HC, $200/ton of CO and $8000/ton of NO_x were based on estimated minimum control costs of other sources in the South Coast region (source: Anderson, 1990). See the discussion in Chapter 2.
6. Information on individual vehicle VMT and emissions was obtained from two vehicle scrappage pilot programmes: Unocal in California and the Illinois EPA in the Chicago area. The vehicles in these samples are older and lower value than average because the buy-back programmes offered under $1000 per vehicle (Unocal offered only $700) and restricted the offers to pre-1975 vehicles (pre-1971 for Unocal). Other data sources provided vehicle emissions (USEPA) or VMT (*RTECS*) but not both.
7. Household demographics and vehicle holdings in the South Coast (source: 1991 CalTrans travel survey of 3200 households in the South Coast region).
8. Emissions savings by model year from scrapping older vehicles (source: California EPA RECLAIM Guidelines).

The question naturally arises as to why the emissions distribution is taken from Indiana rather than California. The one-line answer is that California vehicle emissions data are considered extremely unreliable by the people who know them best.[1] A more technical reason is that California emissions are recorded as a percentage of exhaust flow rather than as grams per mile. The effort to convert such data into the needed grams per mile would be monumental, given that different types of vehicles have different volumes of exhaust flow. Finally, despite stricter new vehicle emissions standards in California, the natural deterioration of emissions systems over time renders the distribution of emissions of older vehicles very similar regardless of location. Vehicles bought in separate scrappage programmes run by Unocal in Southern California and the Illinois EPA (in the vicinity of Hammond, Indiana) had generally similar emissions distributions as Figures C.1–C.3 show. Studies of on-road emissions in the South Coast have also shown that California's emissions are not unlike those of Indiana (Cadle et al., 1993). Table C.1 summarizes all data by source.

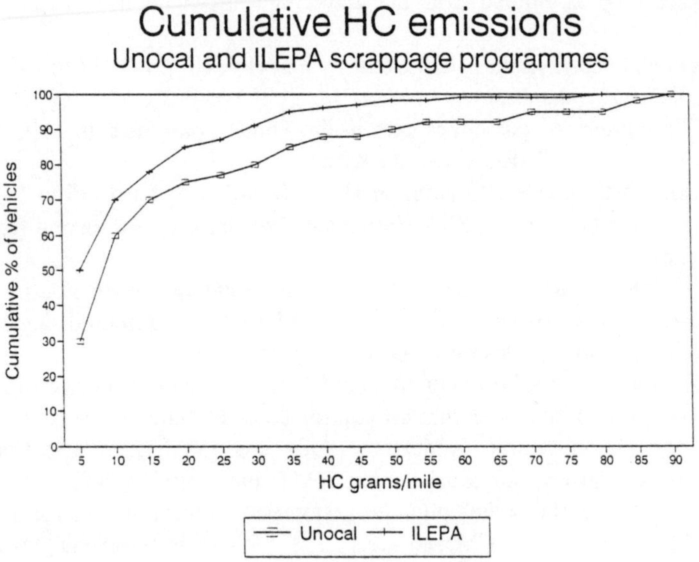

Cumulative HC emissions
Unocal and ILEPA scrappage programmes

Figure C.1 Distribution of HC emissions from vehicles scrapped in California and Illinois pilot programmes

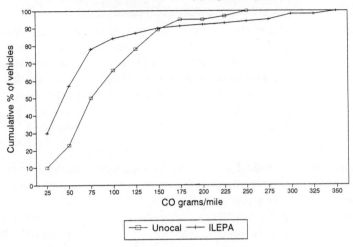

Figure C.2 Distribution of CO emissions from vehicles scrapped in California and Illinois pilot programmes

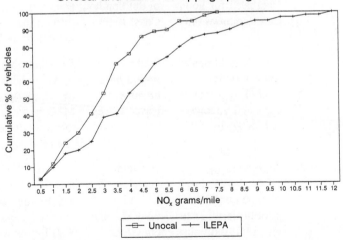

Figure C.3 Distribution of NO$_x$ emissions from vehicles scrapped in California and Illinois pilot programmes

Table C.1 Data sources and description

Source	Type	Sample size	Purpose
CARB	Passenger cars in South Coast region in 1990 and 1995, including number of vehicles, VMT, exhaust and hot and cold start emissions by vintage	Represents all passenger cars in the South Coast region	Calculate annual scrappage rates and emissions deterioration by vintage from 1990 to 1995; use vintage average VMT to calculate VMT changes from the emissions tax
California EPA Air Resources Board	Emissions credits for vehicle scrappage by vintage based on the guidelines from RECLAIM	Not applicable	Estimate emissions savings from induced vehicle scrappage. Credits already adjusted for replacement vehicle and usage
CalTrans	1991 household survey including vehicle holdings and household income. Data used for South Coast counties only	Over 3000 households with over 6700 vehicles in the South Coast region	Analyse household vehicle holdings to estimate impact of the tax on VMT and emissions, including differential impacts by income quintile
CEC	Vehicle holdings and characteristics (including price) by class and vintage (1976–92) in 1992	South Coast region based on state-wide travel survey	Distribution of vehicles by value within vintages for scrappage analysis; MPG by vehicle class; mapping of vehicle classes
Illinois EPA	High-emitting 1968–79 vehicles, scrapped under a pilot project. Data on MPG; individual vehicle HC, CO and NO_x emissions; VMT	207 total vehicles; 112 with data on VMT and emissions	Correlate VMT and HC, VMT and MPG and HC and MPG; compare emissions distribution with vehicles tested in Hammond, Indiana

Source	Type	Sample size	Purpose
USEPA: Hammond, Indiana lane data	I/M 240 readings of HC, CO and NO_x for 1976–92 vintage vehicles in everyday use	Over 10000 vehicles randomly selected	Emissions distribution by vintage for use in TIERS model
US Department of Energy	1991 *Residential Transportation Energy Consumption Survey*. Includes demographics and household vehicle ownership and use	3045 households nation-wide	VMT distribution by model year; vehicle holdings by income quintile; OLS model of VMT for one-, two- and three-or-more-vehicle households; fuel cost in Pacific region
Unocal Corporation	FTP-75 readings of HC, CO, NO_x and MPG for pre-1971 vehicles scrapped for $700	74 vehicles tested out of over 8000 scrapped	Emissions distribution; correlation of emissions with MPG

Appendix D: Description and Flow Chart of the TIERS Model

The TIERS model follows the programming logic described in steps 1–21:

1. Read emissions data from the 1976–91 vehicles tested in Hammond, Indiana. Data record contains vehicle model year, HC grams/mile, CO grams/mile and NO_x grams/mile.
2. Store emissions reading by model year. Emissions are stored in the $HC(obs,yr)$, $CO(obs,yr)$ and $RNO_x(obs,yr)$ arrays. The ith observation of each emission array for the jth model year corresponds to the same vehicle.
3. Allow the user to change the gasoline tax rate. If the purpose of the run is to examine the partial effects of the emissions tax, the gasoline tax should be left unchanged and the user should enter 0.
4. Start the random number generator by obtaining the user's input of the seed value, an odd integer between 1 and 65535.
5. Based on the seed value, randomly choose starting emissions values for each of the 16 model years.
6. Allow user to accept or override the default tax rates of $0.0165/gram of HC, $0.00022/gram of CO and $0.0088/gram of NO_x. User can choose any non-negative rate, including zero, for each pollutant.
7. Allow user to accept or override the default elasticities. This allows the model to test the impact of other studies' price elasticities or perform other sensitivity analyses.
8. Read and store data on vehicle MPG for the 14 vehicle classes in the 16 model years (1976–91). Data are from the CEC's PVMCOND model.
9. Read the average gasoline price per gallon for vehicles in the Pacific region in the *RTECS* sample. Combine data for pre-1976 vehicles with the 1976 vehicles. Because most vehicles use the same grade of gasoline, most vehicles in this sub-sample report the same fuel cost.
10. Calculate the average fuel cost per mile for each class/vintage. Price per mile = (price/gallon)/(miles/gallon).
11. Simulation begins at this point. The user chooses the number of runs to perform. Each run assigns random emissions to each household vehicle in the CalTrans sample, based on the vehicle's model year. The tax per mile

is calculated for these emissions and the increased cost per mile leads to reduced VMT, based on the appropriate elasticities. Reduced VMT, in turn, leads to reduced emissions. Separate subroutines are used to calculate the VMT and emissions reduction for one-vehicle households (subroutine ONEVEH), two-vehicle households (subroutine TWOVEH) and three-or-more-vehicle households (subroutine THREEVEH). The subroutines are described below.

12. Subroutine ONEVEH reads the CalTrans data for one-vehicle households: vehicle type, model year and household demographic weight. Vehicles missing model years (coded as -1) are skipped.

13. ONEVEH calls the subroutine ETAX to draw a random emissions for the vehicle, based on its model year, and calculate the tax per mile. Tax per mile = grams/mile*tax/gram.

14. ONEVEH calls the subroutine BASECOST to calculate the fuel cost per mile for the vehicle. BASECOST maps the vehicle type into one of the 14 vehicle classes or uses the average of the passenger car classes if the vehicle type is missing or classified as other.

15. ONEVEH uses the price elasticity of VMT demand for the one-vehicle household and the percentage change in the cost per mile to calculate the expected percentage change in the vehicle's VMT. (The initial VMT is the average VMT for vehicles of that model year, as reported by CARB.) Initial and adjusted VMTs are multiplied by the household weight to allow cumulation across households.

16. Vehicle emissions prior to the tax are emissions/mile*VMT for each of the three pollutants. Adjusted emissions are calculated as emissions/mile*the adjusted VMT calculated in step 14 above. Initial and adjusted emissions are multiplied by the household weight to allow cumulation across households.

17. VMT, adjusted VMT, initial emissions for each pollutant and adjusted emissions for each pollutant are summed by model year. A grand total for one-vehicle households is also calculated.

18. Tax revenue from one-vehicle households is calculated based on the remaining emissions (in metric tonnes) after the VMT reductions. $Revenue_i$ = annual $emissions_i$*tax $rate_i$ for the ith pollutant. Total revenue is the sum of the revenue received from the three pollutants.

19. Subroutine TWOVEH performs calculations similar to ONEVEH for the two-vehicle households in the CalTrans sample. TWOVEH uses the vehicle's own-price elasticity and the cross-price elasticity (applied to the change in the cost per mile of the other vehicle) to calculate the changes in VMT for each of the household's vehicles. Emissions reductions and tax revenues are calculated based on the VMT reductions, as in ONEVEH. Totals for each model year, and a grand total for two-vehicle households,

are calculated, as in ONEVEH.

20. Subroutine THREEVEH calculates the change in VMT and emissions for each vehicle in households owning three or more vehicles. The cross-price elasticity is applied to the change in the average cost of operating the other household vehicles. Totals for each model year, and a grand total for the three-or-more-vehicle households, are calculated, as in ONEVEH.

21. The main program aggregates the results from ONEVEH, TWOVEH and THREEVEH to report baseline and adjusted VMT and emissions for each model year and the grand total across the entire CalTrans sample.

This ends the first iteration. If the user selected more than one run, the main program zeroes out the VMT and emissions arrays, rewinds the CalTrans household data files, and repeats the simulation (steps 11–21 above) until the required number of runs has been performed. Each run produces detailed and summary reports as described in steps 11–21.

TIERS integrates data and calculations as shown in Figure D.1.

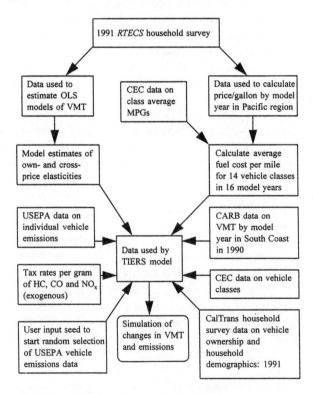

Figure D.1 TIERS model data integration and flow

NOTE

1. For example, a spokesperson at the Bureau of Automotive Repair (BAR), the agency responsible for administering California's automobile emissions inspections, informed the author that she was unable to trace the majority of vehicles through two successive inspection cycles in the BAR database, even though the unique vehicle identification number is supposed to be recorded as part of the inspection report. She suggested that the data were generally unreliable because the persons filling out the inspection reports had no incentive to produce accurate records when the paperwork took time away from their revenue-generating activities. (California's inspections have been administered by private garages.)

Bibliography

American Automobile Manufacturers Association (1993), *AAMA Motor Vehicle Facts & Figures '93*, Washington, DC: American Automobile Manufacturers Association.

Anderson, Robert (1990), *Reducing Emissions from Older Vehicles*, Washington, DC: American Petroleum Institute.

Archibald, Robert and Robert Gillingham (1980), 'An analysis of the short-run consumer demand for gasoline using household survey data', *Review of Economics and Statistics*, **62**, November, 622–8.

Atkinson, Scott E. and Donald H. Lewis (1974), 'A cost-effectiveness analysis of alternative air quality control strategies', *Journal of Environmental Economics and Management*, **1**, 237–50.

Baumol, William J. and Wallace E. Oates (1988), *The Theory of Environmental Policy*, 2nd edn., Cambridge: Cambridge University Press.

Cadle, Steven H., Robert A. Gorse and Douglas R. Lawson (1993), 'Real-world vehicle emissions: a summary of the Third Annual CRC–APRAC On-Road Vehicle Emissions Workshop', *Air & Waste*, **43**, August, 1084–90.

Cadle, Steven H., Mark Carlock, Kevin Cullen, Robert A. Gorse, Kenneth T. Knapp and Douglas R. Lawson (1994), 'Real-world vehicle emissions: a summary of the Fourth Annual CRC–APRAC On-Road Vehicle Emissions Workshop', *Air & Waste*, **44**, October, 1180–87.

California Environmental Protection Agency Air Resources Board (1993), *Mobile Source Emission Reduction Credits: Guidelines for the Generation and Use of Mobile Source Emission Reduction Credits*, prepared by Stationary Source Division and Mobile Source Division, January.

Cambridge Systematics, Inc. (1991), *Transportation Control Measure Information Documents*, draft prepared for the US Environmental Protection Agency, Office of Mobile Sources, Contract No. 68-D9-0073, Work Assignment No. 9, Cambridge, MA, September.

Cambridge Systematics, Inc. (1993), *Positive Feedback Approach to Emissions Reductions for the South Coast Region*, Working Draft – Final Report, prepared for South Coast Air Quality Management District under Contract AB 2766/C0013, Cambridge, MA, November.

Cameron, Michael (1991), *Transportation Efficiency: Tackling Southern California's Air Pollution and Congestion*, Environmental Defense Fund, Regional Institute of Southern California, March.

Dahl, Carol A. (1986), 'Gasoline demand survey', *Energy Journal*, **7**(1), January, 67–82.

Dahl, Carol and Thomas Sterner (1991), 'Analyzing gasoline demand elasticities: a survey', *Energy Economics*, **13**(3), July, 203–10.

Davis, Stacy C. (1994), *Transportation Energy Data Book: Edition 14*, ORNL-6798, prepared for Office of Transportation Technologies, US Department of Energy, Oak Ridge, TN: Oak Ridge National Laboratory, May.

Drollas, Leonidas P. (1984), 'The demand for gasoline: further evidence', *Energy Economics*, **6**(1), January, 71–82.

Ely, E.S. (1994), 'Helping your clunker clear the air', *Business Week*, 7 March, p. 130.

Energy Information Administration, Office of Energy Markets and End Use, US Department of Energy (1993), *Household Vehicles Energy Consumption 1991*, DOE/EIA-0464(91), Washington, DC, December.

Federal Highway Administration (1992a), *Transportation Programs and Provisions of the Clean Air Act Amendments of 1990: A Summary*, publication number FHWA-PD-92-023, Washington, DC: FHWA.

Federal Highway Administration (1992b), *Highway Statistics 1991*, Washington, DC: FHWA.

Findlayson-Pitts, B.J. and J.N. Pitts, Jr (1993), 'Atmospheric chemistry of tropospheric ozone formation: scientific and regulatory implications', *Air & Waste*, **43**, August, 1091–1100.

Golub, Thomas, Seyoung Kim and Weiping Ren (1994), 'A structural model of vehicle use in multi-vehicle households', paper to be presented at the 74th Annual Transportation Research Board Meeting, Washington, DC, 22–28 January, 1995, 7 December.

Goodwin, P.B. (1992), 'A review of new demand elasticities with special reference to short and long run effects of price changes', *Journal of Transport Economics and Policy*, **26**(2), May, 155–64.

Hensher, David A. (1985), 'An econometric model of vehicle use in the household sector', *Transportation Research B*, **19B**(4), 303–13.

Hensher, David A. (1987), 'Automobile loss rates and the expected capital cost of vehicles: an empirical note', *The Economic Record*, September, 247–54.

Hensher, David A., Nariida C. Smith, Frank W. Milthorpe and Peter O. Barnard (1992), *Dimensions of Automobile Demand: A Longitudinal Study of Household Automobile Ownership and Use*, Amsterdam: Elsevier Science Publishers, BV.

Hume, J.N.P. and R.C. Holt (1985), *FORTRAN 77 for Scientists and Engineers*, 2nd edn., Reston, VA: Reston Publishing Co.

Illinois EPA (1993), *Pilot Project for Vehicle Scrappage in Illinois: 1992 Cash for Clunkers*, Springfield, IL: Illinois Environmental Protection Agency, May.

Jack Faucett Associates (1992), *Selected Economic Incentives to Reduce Emissions from Mobile Sources*, final report, submitted to Regulatory Innovations Staff, Office of Policy, Planning, and Evaluation, US Environmental Protection Agency, Bethesda, MD: Jack Faucett Associates, July.

Lawson, Douglas R. (1993), '"Passing the test" – human behavior and California's smog check program', *Air & Waste*, **43**(12), December, 1567–75.

Lents, James M. and William J. Kelly (1993), 'Clearing the air in Los Angeles', *Scientific American*, October, 32–9.

Malecki, Andrew M. (1978), 'Perceived and actual costs of operating cars', *Transportation*, **7**(4), December, 403–15.

Mannering, Fred L. and Clifford Winston (1985), 'A dynamic empirical analysis of household vehicle ownership and utilization', *Rand Journal of Economics*, **16**(2), 215–36.

Maritz Marketing Research, Los Angeles Division (1992), '1991 State-wide travel survey: user's data guide', prepared for Caltrans, California Department of Transportation, 20 February.

Milford, Jana B., Armistead G. Russell and Gregory J. McRae (1989), 'A new approach to photochemical pollution control: implications of spatial patterns in pollutant responses to reductions in nitrogen oxides and reactive organic gas emissions', *Environmental Science and Technology*, **23**(10), 1290–1301.

National Research Council (1991), *Rethinking the Ozone Problem in Urban and Regional Air Pollution*, Washington, DC: National Academy Press.

Parks, Richard W. (1977), 'Determinants of scrapping rates for postwar vintage automobiles', *Econometrica*, **45**(5), July, 1099–1115.

Pisarski, Alan (1992), *1990 Nationwide Personal Transportation Survey: Travel Behavior Issues in the 90s*, prepared for the Office of Highway Information Management HPM-40, US Department of Transportation, Federal Highway Administration, Falls Church, VA: Alan Pisarski, July.

Portney, Paul R. (ed.) (1991), *Public Policies for Environmental Protection*, Washington, DC: Resources for the Future.

Reilly, William J. (1931), *The Law of Retail Gravitation*, New York: W.J. Reilly.

Schere, Kenneth L. (1988), 'Modeling ozone concentrations', *Environmental Science and Technology*, **22**(5), 488–95.

Sigsby, John E. Jr, Silvestre Tejada, William Ray, John M. Lang and John W. Duncan (1987), 'Volatile organic compound emissions from 46 in-use passenger cars', *Environmental Science and Technology*, **21**(5), 446–75.

Sommerville, R.J., Tom Cackette and Thomas C. Austin (1987), 'Evaluation of the California Smog Check Program', SAE Technical Paper Series 870624, International Congress and Exposition, Detroit, MI, 23–27 February.

South Coast Air Quality Management District and Southern California Association of Governments (1991), *1991 Air Quality Management Plan: South Coast Air Basin*, final draft, May.

Stedman, Donald H. (1989), 'Automobile carbon monoxide emission', *Environmental Science and Technology*, **23**(2), 147–9.

Stopher, Peter R. and Arnim H. Meyburg (1975), *Urban Transportation Modeling and Planning*, Lexington, MA: D.C. Heath and Company.

Suits, Daniel B. (1977), 'Measurement of tax progressivity', *American Economic Review*, **67**(5), 747–52.

Train, Kenneth (1986), *Qualitative Choice Analysis: Theory, Econometrics, and Application to Automobile Demand*, Cambridge, MA: MIT Press.

Unocal Corporation (1991), *SCRAP: A Clean-Air Initiative from Unocal*, Los Angeles, CA: Unocal Corporation.

US Congress, Office of Technology Assessment (1992), *Retiring Old Cars: Programs to Save Gasoline and Reduce Emissions*, OTA-E-536, Washington, DC: US Government Printing Office, July.

US Department of Transportation and US Environmental Protection Agency (1993),

Clean Air Through Transportation: Challenges in Meeting National Air Quality Standards, Washington, DC: US DOT and US EPA, August.

Walls, Margaret A., Alan J. Krupnick and H. Carter Hood (1993), *Estimating the Demand for Vehicle-Miles-Traveled Using Household Survey Data: Results from the 1990 Nationwide Personal Transportation Survey*, Discussion Paper ENR-93-25, Washington, DC: Resources for the Future, September.

Wang, Michael Q. and Danilo J. Santini (1994), 'Monetary values of air pollution emissions in various US areas', paper to be presented at the 74th Annual Transportation Research Board Meeting, Washington, DC, 22–28 January, 1995, 15 December.

Zimmer, Barry (1994), 'Work order assignment for "technical support services for planning research"', Contract Number DTFH61-93-C-00075, 8 November.

Index

Air Quality Management Plan (AQMP), 2, 48, 80
American Automobile Manufacturers Association (AAMA), 29, 52, 88n
Anderson, R., 11, 12, 99
AQMP. *see* Air Quality Management Plan
Archibald, R., 34
Atkinson, S.E., 16n

Baumol, W.J., 3, 16n

CAAA. *see* Clean Air Act Amendments of 1990
Cadle, S.H., 16n, 41n, 92, 100
California Air Resources Board. *see* CARB
California Department of Transportation (CalTrans) 1991 household travel survey, 42–44, 90, 99, 102, 104–6
California Energy Commission (CEC), 11, 43, 52, 70, 99, 102, 104
California Environmental Protection Agency, 1, 11, 12, 102
Cambridge Systematics, Inc., 16n, 41n, 93, 94n
Cameron, M., 19, 41n, 55, 94n
CARB, 43, 44, 45, 48, 79, 88n, 90, 99, 102, 105, 106
carbon monoxide, 1, 3, 5, 5n, 6n, 8–12, 17, 43, 45–48, 53–55, 58, 60, 62, 65, 79–80, 87, 88n, 90, 92, 95, 99, 101, 103, 106
Clean Air Act Amendments of 1990, 1–3, 6n, 10, 16n, 90, 97
CO. *see* carbon monoxide

cold start emissions, 10, 66n, 91, 98, 102
'command and control' measures to reduce pollution, 2, 3
congestion externality of travel, 2, 3, 11, 93, 96
cost-effectiveness of emissions-reduction policies 3–4, 8, 89, 94

Dahl, C., 39
Davis, S.C., 88n
Drollas, L.P., 39

elasticity of demand:
 for gasoline, 39
 for VMT, 32, 34, 35, 37, 39, 42
Ely, E.S., 1
emissions tax:
 design of mobile source tax, 5, 7–12, 17, 43
 effect on mobile source emissions, 5, 40, 45–48, 53–55, 58, 60, 62, 64–66
 effect on VMT, 5, 11, 23, 30, 40, 45, 53, 57, 59, 61, 63
 efficiency-enhancing properties, 3–4, 55, 90
 enforcement issues, 10, 92–93
 first best, 5, 8–9
 incidence of, 46, 47, 49–52, 90
 institutional obstacles, 93–94
 rates, 11–12, 99
 revenue-generating potential, 46, 47, 58, 60, 62, 90
 second best, 5, 9–12, 13, 14
 theoretical basis, 3–5, 7, 8–9, 89
enforcement issues of emissions tax, 10, 92–93

112